How Can DoD Compare Damage Costs Against Resilience Investment Costs for Climate-Driven Natural Hazards?

Overview of an Analytic Approach, Its Advantages, and Its Limitations

ANU NARAYANAN, PATRICK MILLS, TOBIAS SYTSMA, KELLY KLIMA, RAHIM ALI

Prepared for the Office of the Secretary of Defense, Office of Cost Assessment and Program Evaluation
Approved for public release; distribution unlimited

RAND NATIONAL DEFENSE RESEARCH INSTITUTE

For more information on this publication, visit **www.rand.org/t/RRA1860-1**.

About RAND

The RAND Corporation is a research organization that develops solutions to public policy challenges to help make communities throughout the world safer and more secure, healthier and more prosperous. RAND is nonprofit, nonpartisan, and committed to the public interest. To learn more about RAND, visit www.rand.org.

Research Integrity

Our mission to help improve policy and decisionmaking through research and analysis is enabled through our core values of quality and objectivity and our unwavering commitment to the highest level of integrity and ethical behavior. To help ensure our research and analysis are rigorous, objective, and nonpartisan, we subject our research publications to a robust and exacting quality-assurance process; avoid both the appearance and reality of financial and other conflicts of interest through staff training, project screening, and a policy of mandatory disclosure; and pursue transparency in our research engagements through our commitment to the open publication of our research findings and recommendations, disclosure of the source of funding of published research, and policies to ensure intellectual independence. For more information, visit www.rand.org/about/principles.

RAND's publications do not necessarily reflect the opinions of its research clients and sponsors.

Library of Congress Cataloging-in-Publication Data is available for this publication.

ISBN: 978-1-9774-1110-5

Cover: Rachelle Blake/U.S. Air Force.

About This Report

Climate change is likely to increase the frequency and severity of extreme weather events, some of which have already affected U.S. Department of Defense (DoD) installations. DoD sets priorities for where and how much to invest in installation resilience to adapt to these hazards. These decisions are complicated by uncertainties of many kinds. It is especially difficult to predict when exactly the next major storm or flood will hit a particular installation. Accounting for this uncertainty in planning decisions can be challenging.

This report focuses on one aspect of this decision calculus—a comparison of the damage costs resulting from extreme weather events against the costs of enhancing installation resilience pre-disaster. This comparison was completed as part of the RAND Corporation's analytic support to the Office of the Secretary of Defense, Office of Cost Assessment Program Evaluation (OSD CAPE) leadership. The final briefing for the project was delivered to the sponsor on June 3, 2022, and the final report was delivered on July 29, 2022.

The research reported here was completed in December 2022 and underwent a security review with the sponsor and the Defense Office of Prepublication and Security Review before public release.

RAND National Security Research Division

This research was sponsored by OSD CAPE and conducted within the Acquisition and Technology Policy Program of the RAND National Security Research Division (NSRD), which operates the National Defense Research Institute (NDRI), a federally funded research and development center sponsored by OSD, the Joint Staff, the Unified Combatant Commands, the Navy, the Marine Corps, the defense agencies, and the defense intelligence enterprise.

For more information on the RAND Acquisition and Technology Policy Program, see www.rand.org/nsrd/atp or contact the director (contact information is provided on the webpage).

Acknowledgments

We would like to thank various people across the Department of the Air Force, the Department of the Navy, and the Department of the Army for providing requisite data for our analysis and for connecting us with points of contact at Langley Air Force Base, Peterson Space Force Base, Cheyenne Mountain Air Force Base, Fort Bragg, and Naval Station Norfolk. Our conversations with these installation personnel helped us contextualize and validate the results obtained through our analysis.

Within RAND, we are grateful to our colleagues Devin Tierney (who helped with the geospatial analysis) and Daniel Kim (who helped with cleaning and coding large amounts of data from the Real Property Assets Database), as well as Chris Mouton and Yun Kang for their program leadership and project guidance. The authors take full responsibility for any remaining errors or inadvertent omissions.

Summary

Issue

U.S. Department of Defense (DoD) installations have been affected by extreme weather events (e.g., Hurricane Michael at Tyndall Air Force Base [AFB], severe flooding at Offutt AFB), and more frequent, less extreme events (e.g., recurrent flooding, hailstorms) that have disrupted missions and resulted in considerable financial loss.[1] DoD needs a way to compare the damage costs resulting from extreme weather events against the costs of mitigating that damage through enhanced installation resilience.[2]

Climate change is likely to increase the frequency and severity of such events, but it is difficult to predict with certainty which installations will be hit and when, or by what sorts of events. It is important for DoD to understand how to account for this uncertainty by setting priorities for where and how much to invest in installation resilience to climate-driven hazards.

Currently, there is no DoD-validated model or method for systematically comparing the costs of damage resulting from extreme weather events against the costs of mitigating that damage by enhancing installations with various resilience options.[3] We begin to address this gap in the report by exploring the relevance and limitations of one analytic approach.

Approach

For each of the 19 case studies—which span military departments, three geographic clusters of installations, and three hazards—we use a simple formula that includes as a parameter the annual frequency with which different hazards are expected to occur to screen for potentially attractive resilience options.[4] We compare the annualized cost of a resilience option with the damage that is averted over that option's lifetime under a wide variety of disaster scenarios. For the illustrative analysis, our focus is on buildings specifically.

First, we estimate damage costs with and without resilience options. We focus on three hazards: coastal flooding, hurricanes (high winds and storm surge), and wildfires. We arrive at the estimates for damage

---

[1] The highest-profile costs are probably those for repairing, recovering, and rebuilding facilities, which receive prominent coverage in the wake of weather events. But installations are subject to many other kinds of costs from extreme weather events, including loss of life and injuries, emergency response, installation operational recovery costs, costs to the mission, and natural resource and environmental costs. Damage of various kinds can be paid for via various funding mechanisms, depending on the size, scope, and type of action. These mechanisms include sustainment, restoration, modernization; military construction; and base operations support. In this report, we do not address the specific funding mechanism that could pay for pre-event resilience options or post-event repair or recovery.

[2] "Resilience is defined for a particular system in terms of particular disruption scenarios. Drawing from the vast literature on the topic, we chose the following definition: '*Resilience* is the ability of a system to withstand and recover from a disruption'" (Anu Narayanan, Debra Knopman, James D. Powers, Bryan Boling, Benjamin M. Miller, Patrick Mills, Kristin Van Abel, Katherine Anania, Blake Cignarella, and Connor P. Jackson, *Air Force Installation Energy Assurance: An Assessment Framework*, RAND Corporation, RR-2066-AF, 2017, p. 6.).

[3] We use the term *resilience option* to apply to any physical change or addition to an installation facility that improves resilience to extreme weather events. The focus of this work is on the facilities or assets (as in real property assets) that make up the installations. Although the presented approach could be applied to a broad set of facilities and assets, we illustrate the approach using just buildings.

[4] For coastal flooding and storm surge, we assess building-level floodproofing and an installation-wide floodwall; for hurricane winds, we assess building-level roofing and hurricane shutters.

resulting from coastal flooding, hurricane winds, and storm surge through simulations run in Hazus, a software tool developed by the Federal Emergency Management Agency. Hazus allows users to run multiple simulations on user-defined building stock data efficiently.

In the absence of a similarly sophisticated approach to estimating damage from wildfire, we found that we would be unable to mirror the method used to compare costs for the other two hazards unless we used the following highly simplified assumptions:

- Any exposure of an asset to wildfire results in the loss of the entire asset.
- Use of the resilience option eliminates exposure to wildfires.

We deemed these assumptions too unrealistic and therefore do not present results for wildfires in this report.

We then compared what we term the *averted damage* against the costs of implementing the resilience options (estimated using a cost model we developed) for a variety of frequencies with which hazard events of varying severities might occur. For flooding and hurricane winds, the difference between the damage without and with resilience options in place is the averted damage or the damage cost differential. The result is an estimate of the increased frequency (compared with a baseline of the historical record) with which hazard events would need to occur before the averted damage exceeds the cost of each relevant resilience option for each hazard of interest.

Table S.1 shows the case studies to which we applied these methods. The selected case studies include installations that span military departments and focus on the three types of hazards (coastal flooding, hurricanes [high winds and storm surge], and wildfires). As we just mentioned, lack of data and appropriate modeling tools for assessing installation wildfire exposure required us to make unrealistic assumptions to conduct an analysis that matches the structure used for the other hazards. Therefore, we do not report results for the wildfire analysis in this report, but we do describe our analytical attempt and underlying.

The presented analysis has limitations. Some of the limitations stem from gaps in data availability, methods, or tools, which present potential opportunities for DoD to act to close these gaps. Another group of limitations results from choices we made during the course of this analysis, mostly because of limited project time and resources. These limitations could potentially be addressed by future research.

TABLE S.1
Case Studies

Cluster	Installation	Military Department	Coastal Flooding	Hurricane Winds	Storm Surge	Wildfire[a]
Norfolk (Va.)	Naval Station Norfolk (plus NAVSUPPACT Hampton Roads and NAVSUPPACT Norfolk Naval Shipyard)	Navy	X	X	X	
	Joint Base Langley-Eustis	Air Force, Army	X	X	X	
	Naval Weapons Station Yorktown	Navy	X	X	X	
Fort Bragg (N.C.)	Fort Bragg	Army		X		
Colorado	Peterson SFB	Air Force				X
	Schriever SFB	Air Force				X
	Fort Carson	Army				X

NOTE: *Cluster* refers to groups of geographically close installations. Blank spaces mean that the installation does not experience the hazard because of its geographic location. NAVSUPPACT = naval support activity; SFB = space force base.

[a] We include the wildfire case studies here for completeness but do not discuss this hazard in the rest of the report.

Key Takeaways

- We demonstrate that Hazus or a similar tool could be used to begin to understand the value of investing in installation resilience to climate-driven hazards.
- That said, the availability of data to support analyses of the sort we present in this report is limited.
  - Although central DoD databases are a credible place to take stock of asset inventories at installations (numbers and types of assets), no single database offers comprehensive, high-quality data related to asset-level features, such as geolocations, plant replacement value, other measures of financial value, and resilience options already in place.
  - Available data and tools to characterize the wildfire hazard fall considerably short compared with those that are available for flooding and high wind hazards.
- This report identifies data availability and gaps, both within and outside DoD, that can serve the methods we demonstrate.
- The results presented in this report are illustrative, and we do not recommend they be used directly to inform resilience investment decisions. Rather, they are intended to guide additional data collection and analysis that can highlight specific courses of action DoD should take to improve resilience to climate change.
- One alternative to the modeling and simulation–based analysis we present in this report is to learn from real-world events as they occur and from historical installation storm damage data. But this approach has downsides: Data might not have been captured for some installations, data might be difficult to extract from installation sources, and many DoD installations might not have been affected by enough extreme weather events to have sufficient data points. For example, assumptions might need to be made about the way in which a storm of a given type would affect facilities at location B, having occurred only at location A.

Recommended Options for a Way Forward

There are three broad ways in which the analysis presented in this report could be furthered.

- At the enterprise level (DoD, Office of the Secretary of Defense [OSD], or military service), the Hazus model, the resilience option cost data, and the other non-DoD data and methods used in our approach need to be validated, verified, and customized for DoD purposes before being used in specific decision-making contexts. Alternatively, investment could be made to develop different tools for DoD purposes.
- Furthermore, enterprise-level analysis would require augmenting centrally available data with installation-provided data to perform the analysis once the methods have been validated and customized for DoD installations.
- Decentralized, installation-level analysis would require key enablers, such as guidance on how to evaluate resilience options; access to, and familiarity with, analytic tools and how to apply and adapt them for DoD installation purposes (such as using Hazus); and mechanisms to track and record storm damage data.

Regardless of which of these approaches is pursued, OSD could play a role by issuing broad guidance to the Services about *how* to evaluate resilience options; by taking steps to improve the quality of the Real Property Assets Database and of installation geospatial data; by establishing standard mechanisms through which installations can track, record, and report data related to storm damage; or by helping improve the familiarity of Service- or DoD-level personnel with analytical tools for estimating disaster-related damage.

Contents

Figures and Tables

Figures

Tables

CHAPTER 1

Introduction

The U.S. Department of Defense (DoD) is responsible for more than 585,000 facilities located on 4,775 sites worldwide, with an estimated value of more than $1.17 billion.[1] These sites can be small outposts (e.g., satellite monitoring stations, recruiting stations) or sprawling bases with tens of thousands of personnel, massive industrial operations, and state-of-the-art weapon systems.

The installations that contain these sites (a single installation can have accountability for multiple, sometimes geographically separated, sites) have a diverse array of facility types and perform a variety of missions. DoD must set policy and manage resources for the facilities at these many disparate installations in a way that conserves funding and meets mission needs. Allocating resources and managing this enormous portfolio of installations and facilities is already challenging in the typical budget environment and in a permissive operating environment.[2]

However, extreme weather events have been increasing since the 1980s (or since 2000, depending on how you count), and DoD installations have been affected by several very high-profile cases since at least 2010 that have caused billions of dollars in damage and disrupted and degraded missions.[3] Notably, although they are not as newsworthy, more-frequent but less-extreme events (e.g., recurrent flooding, hailstorms) at installations have disrupted missions and resulted in considerable financial loss. Climate change is likely to increase the frequency and severity of such events, but it is difficult to predict with certainty which installations will be hit, when they might be hit, and by what sorts of events.[4]

DoD has several responsibilities related to climate-driven disruptions to installation performance and missions. One such responsibility is to design policies that help the Services cope with the damage and disruption from extreme weather events. To do this, DoD needs to account for the potential damage and disrup-

---

[1] DoD, *Base Structure Report–Fiscal Year 2018 Baseline: A Summary of the Real Property Inventory Data*, 2018. DoD Instruction (DoDI) 4165.14 defines a *facility* as

> a building, structure, or linear structure whose footprint extends to an imaginary line surrounding a facility at a distance of 5 feet from the foundation that, barring specific direction to the contrary such as a utility privatization agreement, denotes what is included in the basic record for the facility (e.g., landscaping, sidewalks, utility connections). A facility will have a Real Property Unique Identifier (RPUID) and is entered into a Service [Real Property Inventory] system as a unique real property record. (Department of Defense Instruction No. 4165.14, *Real Property Inventory (RPI) and Forecasting*, Department of Defense, January 17, 2014, incorporating change 2, August 31, 2018)

[2] DoD installations are subject to many different threats and hazards, including natural and man-made. Our focus here is only on extreme weather events.

[3] Some examples are Hurricane Florence, which struck Camp Lejeune, North Carolina, in September 2018; Hurricane Michael at Tyndall Air Force Base (AFB) in October 2018; severe flooding at Naval Support Activity Mid-South in Millington, Tennessee, in 2010; and Winter Storm Uri in February 2021, which caused significant damage in Texas.

[4] Tom Knutson, "Global Warming and Hurricanes: An Overview of Current Research Results," Geophysical Fluid Dynamics Laboratory, last updated February 9, 2023; Department of Defense Strategic Environmental Research and Development Program and Environmental Security Technology Certification Program, "Regionalized Sea Level Change Scenarios," database, undated; Rebecca Lindsey, "Climate Change: Global Sea Level," National Oceanic and Atmospheric Administration Climate.gov, April 19, 2022.

tion from climate-driven hazards and especially for future climate uncertainty. DoD has taken steps in recent years to better characterize the projected exposure of installations to various climate-driven hazards (e.g., the Defense Climate Assessment Tool).

Such exposure data serve as the backdrop for the analysis we present in this report, which aims to facilitate comparisons of the financial implications of investing in resilience with climate-driven hazards.[5] At present, there is no DoD-validated model or method for systematically comparing the costs of damage resulting from climate-driven hazards against the costs of improving resilience to those hazards. This report begins to address this gap by exploring the relevance and limitations of one analytic approach.

Our approach draws on DoD and commercial data and uses Hazus, a software tool developed by the Federal Emergency Management Agency (FEMA) that efficiently runs multiple simulations on user-defined building stock data. Although Hazus has not been extensively used in the DoD context and cannot tackle the diversity of DoD installation buildings (let alone the full spectrum of assets that compose DoD installations), others have used the tool to understand risks to military installations.[6]

Mission Accomplishment and Fiscal Responsibility in Decisionmaking

DoD must consider, and sometimes balance, numerous factors in its decisionmaking, chief among these being mission accomplishment and financial cost. Although DoD must operate cost-effectively, the mission comes first. Installations support missions, either directly via mission facilities, or indirectly via the variety of support facilities that provide needed services to Service members and their families (e.g., utilities, security).

Extreme weather events can disrupt missions in two ways. First, they can disrupt missions directly, diverting mission resources to prepare for and recover from such events. Second, damage to installation facilities can prolong the mission disruptions that result from disasters. Resilience investments and changes to operating procedures could reduce these disruptions, thereby preserving mission capability and capacity.[7]

Likewise, financial considerations drive decisionmaking. Disaster recovery costs can be considerable and investing now in more-resilient facilities might reduce future expenses for the Services, DoD, and taxpayers. Other objectives for installations include environmental compliance and stewardship, as well as quality of life goals for Service members and their families. These goals also factor into Service-level and local decisionmaking about installations.

---

[5] For ease of analysis, we use exposure data contained within a model called Hazus for the hurricane winds and flooding analysis instead of the Defense Climate Assessment Tool. Given our reliance on Hazus to estimate incurred damage, it made sense to also use embedded hurricane wind speed and flood exposure data to conduct our analysis. We describe the Hazus model and its strengths and limitations in Chapter 2.

[6] See, for instance, Alexander J. Baldwin, *Developing Infrastructure Adaptation Pathways to Combat Hurricane Intensification: A Coupled Storm Simulation and Economic Modeling Framework for Coastal Installations*, thesis, Air Force Institute of Technology, March 2021; Kiara L. Vance, *Natural Infrastructure Alternatives Mitigate Hurricane-Driven Flood Vulnerability: Application to Tyndall Air Force Base*, thesis, Air Force Institute of Technology, March 2022.

[7] There are several definitions of resilience, but most are variations of "'the ability to withstand and recover from a disruption'" (Anu Narayanan, Debra Knopman, James D. Powers, Bryan Boling, Benjamin M. Miller, Patrick Mills, Kristin Van Abel, Katherine Anania, Blake Cignarella, and Connor P. Jackson, *Air Force Installation Energy Assurance: An Assessment Framework*, RAND Corporation, RR-2066-AF, 2017, p. 6.) For example, one source states that

> [m]ilitary installation resilience is defined as the capability of a military installation to avoid, prepare for, minimize the effect of, adapt to, and recover from extreme weather events, or from anticipated or unanticipated changes in environmental conditions, that do, or have the potential to, adversely affect the military installation or essential transportation, logistical, or other necessary resources outside the military installation that are necessary to maintain, improve, or rapidly reestablish installation mission assurance and mission-essential functions. (Office of Local Defense Community Cooperation, "Installation Resilience," webpage, undated)

Acknowledging the diverse set of goals that DoD must balance when making decisions regarding where and how much to spend on infrastructure resilience, we focus in this report on the financial aspect.

Research Questions

At the outset of this project, the sponsor posed a high-level research question: How do the costs of investments in resilience to climate-driven extreme weather events compare with post-disaster damage costs?

To address this question, we designed our analysis to answer four underlying questions, which are the focus of this report.

- **Is there a systematic and repeatable way to answer this question?** Currently, there is no DoD-validated model or method for systematically assessing and comparing the relevant costs. This report begins to address this gap by exploring the relevance and limitations of one analytic approach. Two additional questions stem from this first question.
- **What types of data, information, methods, and tools are available to answer this question?** As this report will show, the breadth and diversity of data required to feed the kinds of scalable, repeatable methods we demonstrate here can be daunting. We sought to assess what data are available, both within and outside DoD, to serve the methods we demonstrate in this report, and how much effort is required to prepare them.
- **Which hazards, facility types, and resilience options lend themselves to the type of analysis sought?** Another issue we examine is the suitability of this type of analysis for assessing the comparative costs of addressing various hazards. We attempt to demonstrate the attributes of hazards, facility types, and resilience options that permit the kind of analysis we undertook. The findings along these lines are not exhaustive (in Chapter 2, we explain the bounds of our scope), but they should be instructive.
- **What can DoD do to further the state of the art in performing these kinds of analyses?** Our recommendations focus on helping DoD support this kind of analysis at the enterprise and installation levels.

Limitations of This Analysis

Notwithstanding the questions this analysis can inform, there are several important limitations. Some of these analytic limitations were because of corresponding limitations in the availability of satisfactory data, methods, or tools. These offer potential opportunities for DoD to act to close the gaps. Another group of limitations results from choices we made during the course of this analysis, mostly because of limited project time and resources. These limitations could be potentially addressed by future research. Table 1.1 summarizes these limitations. Later chapters explain each in more detail.

We do not recommend either a proactive or reactive approach as the most appropriate policy choice across contexts. The million-dollar question, so to speak, is whether it pays off financially to invest in resilience before disaster strikes or to simply pay for recovery afterward (despite potential mission consequences). What we found, and what intuition already tells us, is that the so-called payoff depends on many different factors, some of which are sensitive to time, some to very specific local conditions, and so on. Even an exhaustive version of the analysis we present would not necessarily pick winners in such a way as to definitively prescribe particular investments as being universally cost-effective.

Finally, although the analytic *approach* is applicable to the broad set of DoD installations and to the hazards we cover in this report, the *results* for the selected case studies should not be extrapolated to derive insights regarding other installations. For instance, just because installation A might look similar in compo-

TABLE 1.1

Summary of Key Limitations by Category

Limitation	Implication
Data, Methods, Tools	
Hazus has not been validated for DoD installations or facilities	Limits immediate applicability of quantitative results to investment decisions
Hazus does not capture full diversity of DoD facilities and only generically (using census data) captures existing features; best equipped to handle certain types of buildings[a]	Illustrative analysis does not capture the full damage costs of hazards because only buildings are considered
	A rough map of Hazus building types to DoD building types (i.e., every DoD building type did not have a perfect match in Hazus) involves subjective judgments
	For buildings included in the analysis, damage costs might be overstated because any resilience options already in place are not explicitly considered
Hazus does not include all possible resilience options to address hazards	Potentially available resilience options might offer better trade-offs than those presented here
There is a lack of comprehensive data on the value of building contents	Could understate benefits of resilience options
There is not a sufficiently sophisticated approach to estimating damage from wildfires	Could not do credible wildfire damage or trade-off analysis
Hazard-specific limitations relate to level of uncertainty (e.g., storm surge analysis maps Hazus wind speeds to the National Oceanic and Atmospheric Administration's [NOAA] Sea, Lake, and Overland Surges from Hurricanes [SLOSH] model outputs, which are not perfectly aligned)[b]	Uncertainty of hazard frequency estimates and associated damage to varying degrees across hazards
Project Scope	
The costs of ongoing maintenance or repairs were not considered	Could overstate the net cost of resilience options
The potential increase in maintenance costs from the resilience options were not considered	Could understate the annualized cost of implementing each option
How resilience implementation would be financed was not investigated, and how the period over which a resilience option investment is repaid was only minimally investigated, which might change the results	Could alter when and where resilience implementation yields cost savings
Only a small set of resilience options for each hazard were compared	Options are potentially available that would offer better trade-offs than those presented here
The model runs did not include the full scope of installation-level investment decision factors, such as mission considerations or budget constraints (e.g., we assume that resilience options are applied to all buildings, not just those that might be viewed as mission critical; limiting the pool of buildings that receive resilience upgrades would change the results)	Limits immediate applicability of quantitative results to investment decisions
Sensitivity analyses of the modeling results to explore uncertainties in the resilience option cost factors or PRV were not conducted	Cost differences could alter trade-offs of different resilience options

Limitation	Implication

NOTE: PRV = plant replacement value.

[a] As defined by DoDI 4165.14 (2018, p. 15), a building is "a roofed and floored facility enclosed by exterior walls and consisting of one or more levels that is suitable for single or multiple functions and that protects human beings and their properties from direct harsh effects of weather such as rain, wind, sun, etc." Buildings are one of three types of facilities in DoD, with the two other types being *linear structures* (a facility that has a required function to traverse land or is otherwise managed or reported by a linear unit of measure, e.g., runways) and *structures* (a facility, other than a building or a linear structure, that is constructed on or in the land). For instance, Hazus does not specifically include larger-sized buildings that are common in the military context (e.g., maintenance buildings to repair larger combat vehicles).

[b] NOAA, "SLOSH Display Package," webpage, undated.

sition (in terms of number or types of assets) to our analysis of installation B does not mean that our results would look similar for installation A for a given hazard. Factors such as data availability (which governs which assets are analyzable for a given installation) and local conditions (which govern how a given hazard affects assets) heavily influence the answer. We discuss limitations specific to the data and the models needed to carry out different elements of the analysis in the respective sections of Chapter 2. We discuss other limitations in more detail in the final section of Chapter 3 and in the "Key Findings" section of Chapter 5.

Organization of This Report

Chapter 2 provides an overview of the underlying method and data sources used, including the case studies we selected.

Chapter 3 presents the results of the analysis, including estimated damage costs to facilities with and without proactive resilience options in place, the cost of those resilience options, and comparisons of averted damage costs with the resilience option costs.

Chapter 4 highlights the importance of considering the mission and other implications of investments in resilience. Though it is out of the scope of this report to treat this topic in greater depth, we provide this discussion to add context to our financially focused analysis.

Chapter 5 presents a summary of the results; findings related to the analytic approach, data needs, and data availability; and recommended options for DoD going forward.

Finally, we present three appendixes with technical details. Appendix A documents Hazus, the primary model we used. Appendix B documents our approach to analyzing resilience improvement versus repairing damage from wildfires. Appendix C presents our cost model for resilience options.

CHAPTER 2

Methods and Data

The decision of whether to invest in improving installation resilience to extreme weather events involves a significant amount of uncertainty. Not investing in resilience options certainly has downsides. Post-disaster damage costs are often high and are likely to grow as the frequency or severity of extreme weather events increases with climate change. However, resilience options can be costly and might pay off only if an event occurs. It is also difficult to accurately predict when the next disaster will strike a particular location. Some of these uncertainties are knowable. For example, the period over which a resilience option is financed is knowable but will vary from case to case. Other uncertainties are generally unknowable: For example, some disasters are likely to become more frequent or intense over the next century because of climate change, but neither the exact magnitude of these changes nor the specific installations that will be affected by them can be known with certainty.

The frequency with which hazard events of interest occur over a given investment's lifetime is a key parameter that governs whether the benefits of the investment outweigh its costs. It is also one of the most significant sources of uncertainty surrounding when and where to invest in resilience. For example, for a given installation, the averted damage could outweigh the investment cost only when a higher-severity disaster occurs but not for milder events that might occur more frequently. Additionally, an investment might pay off only under scenarios where events occur much more frequently than they do currently. Understanding the relationship between the costs and benefits of an investment, and the probabilistic frequency and severity of disasters, is critical in the context of climate change, where disasters are projected to become more frequent or intense over the next century.[1]

Approach

In this report, we use case studies (selected in coordination with the sponsor and described in detail in a subsequent section of this chapter) to illustrate an approach for comparing the costs of investments in installation resilience against the damage costs (to buildings) that result from extreme weather events for a select set of event types that are anticipated to worsen with climate change. The selected case studies include installations that span military departments and focus on three types of hazards: coastal flooding, hurricanes (high winds and storm surge), and wildfires. For each type of hazard, the method is geared toward lower-frequency and higher-severity events (events with return periods of at least a few decades, as opposed to nuisance events that occur frequently).[2] But nothing inherent to the structure of the analysis prevents it from being used to examine the implications of events that are less extreme or more frequent.

---

[1] Donald J. Wuebbles, David W. Fahey, and Kathy A. Hibbard, eds., *Climate Science Special Report: Fourth National Climate Assessment*, Volume I, U.S. Global Change Research Program, 2017.

[2] The *return period* is an estimate of the time between events of similar magnitude or severity. A 50-year return period wind event at two locations can be associated with two different wind speeds. Although a return period for an event (e.g., 50 years)

For each {installation, hazard} pair, we use a simple formula that includes as a parameter the annual frequency with which different hazards are expected to occur to screen for potentially attractive resilience options. We compare the annualized cost of a resilience option with the damage it averts over its lifetime under a wide variety of disaster scenarios. First, we estimate damage costs with and without resilience options. For coastal flooding and hurricane winds and storm surge, we arrive at these estimates through simulations run in Hazus, a software tool developed by FEMA (with coastal flood depths from the Strategic Environmental Research and Development Program [SERDP] and the Environmental Security Technology Certification Program [ESTCP]).[3] Hazus provides estimates of the extent of degradation that assets might experience when subjected to hazards of different severities. We then multiply these Hazus outputs by the PRV to estimate damage costs. Although the presented approach could be used to compare damage versus resilience costs for any set of assets or facilities, assuming requisite data are available, we focus our analysis on buildings.

For wildfires, in the absence of a similarly sophisticated approach to estimating damage, we found that we would be unable to mirror the use of Hazus:

- Any exposure of an asset to wildfire results in the loss of the entire asset.
- Use of the resilience option eliminates exposure to wildfires.

We deemed these assumptions too unrealistic. Although we document our approach to addressing the wildfire hazard in Appendix B, we will not discuss this hazard further in this report and will instead focus on the high winds of hurricanes and flooding hazards.

For the flooding and hurricane winds hazards, the difference between the damage without and with resilience options in place is the damage cost differential or the averted damage provided by the resilience option. First, we calculate the savings from the resilience option by comparing the differential against the option's cost. We then assess how these savings change under future climate conditions, which we simulate by increasing the frequency with which hazard events of a specified severity occur.

For each resilience option, installation, and hazard type, we calculate the annualized cost of averted damage as

$$AvertedDamage = \sum_{\{i=1\}}^{m} \frac{\sum_{\{n=1\}}^{L}\left[\binom{L}{n}*EA\,P_i^{n}*(1 - EA\,P_i)^{L-n}\right]*\left[BaseDamage - ResDamage\right]}{L}, \tag{1}$$

where

- m is the number of event return periods (or severities) for which we have data
- L is the lifetime of the resilience option
- n is the number of events of a given severity

---

could serve as a proxy for a specific element of severity (e.g., 65-mph wind speed) at a given location, it cannot generically serve as a proxy for event severities at *all* locations. We use return period and severity interchangeably in this report when discussing severity in the context of a specific location.

[3] Hazus is a computer program developed by FEMA that allows users to simulate natural disasters and evaluate damage under various conditions. The program allows the user great flexibility in scenario building to calculate damage from events of different magnitudes and return periods. Hazus could, for example, incorporate such information as hurricane probabilities and digital elevation maps (DEMs) to define a hurricane's wind and coastal flooding damage. Hazus allows users to run multiple simulations on user-defined building stock data efficiently and has been used in previous analyses of DoD installation disaster damage (FEMA, *Hazus 5.1 Flood Model User Guidance*, April 2022a; FEMA, *Hazus 5.1 Hurricane Model User Guidance*, April 2022b).

- *EAP*$_i$ is the expected annual probability of an event of a given severity occurring (as presented by the *i*th of *m* event severities included in the analysis)
- *BaseDamage* is the damage cost in the baseline without any additional resilience option in place
- *ResDamage* is the damage cost when a resilience option is used.

In the numerator, the term

$$\left[\binom{L}{n} * EAP_i^n * \left(1 - EAP_i\right)^{L-n} \right]$$

is the probability of *n* events of a given severity (represented by the event return period) occurring within a given year for each of the *L* years of the option's lifetime. For instance, the *expected annual probability* (EAP) of a 50-year flood event is 0.02 (or one in 50). The second term in the numerator is the difference in damage without and with the resilience option each time an event of a given severity occurs. The product in the numerator is summed over the lifetime of the resilience option (*L*), and then divided by the lifetime of the resilience option.[4] Finally, the outside summation calculates all event return periods (*m*) for the location for which we were credibly able to estimate damage using the data and tools available to us.

We compare the costs of averted damage calculated in the equation above to the annualized cost of implementing the resilience option. The result is the annual savings for the resilience option. When the averaged damage costs exceed the cost of the resilience option, the annual savings are positive and the resilience option shows promise. However, when the averted damage is less than the resilience costs, the resilience option could still show promise if the expected annual probability of events changes because of climate change. The equation allows us to simulate how much more frequent natural disasters would need to become for each resilience option we considered to pay off.

In the following sections, we discuss the data used in the analysis, our methods for selecting installations and hazards for the case studies, how we process the data, and several limitations of our analysis.

Data

We used a variety of data sources to identify relevant installations and assets, estimate hazard damage, and estimate the costs of resilience options. Table 2.1 summarizes these data sources. Data on hurricane winds and coastal flood depths come primarily from Hazus. We discuss Hazus in greater detail below and in Appendix A. Hurricane storm surge depth maps come from the National Hurricane Center (see the "Step 3c: Storm Surge" section in Appendix A for additional details).[5] We used wildfire burn probability data from the U.S. Forest Service (USFS) (see Appendix B for additional details) to explore the feasibility of a wildfire analysis,[6] but as noted previously, we ultimately do not present results from the wildfire analysis in Chapter 3.

The main source of data on installation inventory comes from the Real Property Assets Database (RPAD).[7] This dataset provides information on the PRV, age, area, number of stories, and descriptions of all assets at an

---

[4] In Chapter 3, we discuss a couple of excursions that vary the period over which the resilience option cost is annualized to show the effect of this assumption on results.

[5] National Hurricane Center and Central Pacific Hurricane Center, "Storm Surge Maximum of the Maximum (MOM)," webpage, undated-b.

[6] Karen C. Short, Mark A. Finney, Joe H. Scott, Julie W. Gilbertson-Day, and Isaac C. Grenfell, "Spatial Dataset of Probabilistic Wildfire Risk Components for the Conterminous United States," 1st ed., Forest Service Research Data Archive, 2016.

[7] RPAD is a consolidated summary of the military departments' native Real Property Inventory (RPI) data. RPAD is the recognized source for DoD RPI data reporting requirements throughout the year, including the Federal Real Property Profile

TABLE 2.1
Data Sources

Data Type	Source	Description
Hazards		
Hurricane winds	Hazus hurricane wind model	Hurricane wind speeds based on probabilistic hurricane scenarios in Hazus
Coastal floods	Hazus coastal flood model	Flood depth based on probabilistic coastal flood scenarios in Hazus, with coastal flood depths from SERDP and ESTCP
Storm surge	National Hurricane Center	Flood depth from hurricane storm surge based on Maximum of Maximum Envelope of Open Water maps
Installation Data		
Installation inventory	RPAD	PRV, building age, and building area for individual assets within an installation
Asset locations	GIS	Geographic locations for individual assets within an installation
Cost Data		
Building damage costs	Hazus, RPAD	PRV from RPAD as proxy for value of assets included in the analysis, multiplied by Hazus outputs of asset degradation (as a percentage of asset value) resulting from exposure to hazards of varying severities
Resilience option costs	RSMeans, literature	Data on unit costs of resilience options

NOTE: GIS = geographic information system.

installation. The PRV, which is calculated using several DoD cost factors, plays a key role in our calculation of damage costs.[8]

However, critically, RPAD does not include data on the geolocation of assets. For this, we turn to GIS data provided by Department of the Army G-9 Army Environmental Division, the Naval Facilities Engineering Systems Command Mid-Atlantic, and the Air Force Civil Engineer Center. We discuss how we merge RPAD data with GIS data in Appendix A.

Finally, we were unable to find a comprehensive, cross-Service data source that provided cost factors for the sorts of resilience options we include in this analysis. Therefore, data used to estimate resilience option costs come from several sources. We used unit cost data from RSMeans, a construction estimating software, to estimate the costs for several resilience options.[9] Where we found RSMeans data lacking, we relied on resilience cost estimates from the natural hazard resilience literature. We discuss these data sources in greater detail in Appendix C. We informally compared our estimates of resilience option costs for one or two options

---

and National Real Property Efficiency reporting summaries (DoD, 2018). We obtained RPAD data for this project from the Office of the Secretary of Defense, Office of Cost Assessment and Program Evaluation (OSD CAPE) on February 25, 2022.

[8] The standard DoD formula for calculating PRV is as follows: Plant Replacement Value = Facility Quantity × Replacement Unit Cost × Area Cost Factor × Historical Records Adjustment × Planning and Design Factor × Supervision Inspection and Overhead Factor × Contingency Factor (DoD, *Unified Facilities Criteria (UFC): DoD Facilities Pricing Guide*, UFC 3-701-01, March 17, 2022, incorporating change 1, July 15, 2022a).

[9] Gordian, "RSMeans Data," webpage, undated.

with those provided by DoD subject-matter experts, but time constraints kept us from undertaking such a systematic comparison for each considered option.[10]

Case Study Selection

Our analysis focuses on hazards that are likely to increase in frequency and severity with climate change. Numerical simulation models indicate that the number of very intense hurricanes is likely to increase in a warmer climate, that coastal flood events are expected to grow in frequency and severity because of sea level rise, and very large wildfires are likely to become significantly more common in large parts of the United States.[11] These trends led us to focus our analysis on hurricane winds and storm surge, coastal floods, and wildfires. The mid-Atlantic seaboard is exposed to both hurricanes and coastal flooding events.[12] The western part of the United States, including the Rocky Mountain region, is exposed to wildfire.[13]

We selected installations that (1) are in these hazard-exposed regions, (2) have high PRV, and (3) cover all three military departments. To incorporate a broader variety of installations, we created *clusters*, groups of geographically close installations that, together, have high PRV. Specifically, we included five installations in the Norfolk, Virginia, region, which we refer to as the Norfolk cluster. Additionally, we included three installations in Colorado, which we refer to as the Colorado cluster. Finally, we include Fort Bragg by itself, given this installation's high PRV. Table 2.2 displays the results of the hazard and installation screening processes.

Within installations, we further screen the assets included in the analysis. Specifically, we focus on assets that (1) have a high replacement value, (2) could be matched with GIS data, (3) are likely susceptible to hazards, and (4) are compatible with Hazus. Note, the last criterion pertains only to floods and hurricanes, because the wildfire analysis did not involve Hazus. This screening process results in an analysis that focuses on buildings, a broad category of assets that includes hospitals, military housing, offices, storage facilities, and other assets with walls and a roof. Buildings account for 68 percent of total property PRV across all installations, making them relevant in terms of costs.[14] Additionally, buildings are easily matched with GIS

---

[10] We spoke with approximately 15 personnel spanning five installations—Fort Bragg, Naval Station Norfolk, Langley AFB, Peterson Space Force Base, and Cheyenne Mountain AFB—and Service organizations with regional or enterprise-wide responsibilities from Naval Facilities Engineering Systems Command Mid-Atlantic Region; Office of the Assistant Secretary of the Army (Installations, Energy, and Environment); Department of the Army G-9 Army Environmental Division; U.S. Army Installation Management Command; and U.S. Air Force Energy, Installations, and Environment. The primary purpose of these conversations was to ground our analysis and to ensure that the installation perspective informed our work, not to gather information that would directly feed into our analysis.

[11] James P. Kossin, Timothy Hall, Thomas Knutson, Kenneth E. Kunkel, Robert J. Trapp, Duane E. Waliser, and Michael F. Wehner, "Extreme Storms," in Donald J. Wuebbles, David W. Fahey, and Kathy A. Hibbard, eds., *Climate Science Special Report: Fourth National Climate Assessment*, Volume I, U.S. Global Change Research Program, 2017, pp. 257–276; William V. Sweet, Radley Horton, Robert E. Kopp, Allegra N. LeGrande, and Anastasia Romanou, "Sea Level Rise," in Donald J. Wuebbles, David W. Fahey, and Kathy A. Hibbard, eds., *Climate Science Special Report: Fourth National Climate Assessment*, Volume I, U.S. Global Change Research Program, 2017, pp. 333–363; E. Natasha Stavros, John T. Abatzoglou, Donald McKenzie, and Narasimhan K. Larkin, "Regional Projections of the Likelihood of Very Large Wildland Fires Under a Changing Climate in the Contiguous Western United States," *Climatic Change*, Vol. 126, 2014.

[12] Wuebbles et al., 2017.

[13] Stavros et al., 2014.

[14] We use the general term *military housing* to refer to several types of housing represented in the data, including family housing dwelling (which constitutes the largest category of housing), enlisted unaccompanied personnel housing, transient lodging service academy, and unaccompanied personnel housing. *Total property PRV* refers to the value of all real property, including buildings and non-building structures, such as roads, runways, and utilities. It does not include the cost of equipment inside buildings.

TABLE 2.2
Case Studies

Cluster	Installation	Military Department	Coastal Flooding	Hurricane Winds	Storm Surge	Wildfire[a]
Norfolk (Va.)	Naval Station Norfolk (plus NAVSUPPACT Hampton Roads and NAVSUPPACT Norfolk NSY)	Navy	X	X	X	
	Joint Base Langley-Eustis	Air Force, Army	X	X	X	
	Naval Weapons Station Yorktown	Navy	X	X	X	
Fort Bragg (N.C.)	Fort Bragg	Army		X		
Colorado	Peterson SFB	Air Force				X
	Schriever SFB	Air Force				X
	Fort Carson	Army				X

NOTE: *Cluster* refers to groups of geographically close installations. Blank spaces mean that the installation does not experience the hazard because of its geographic location. NAVSUPPACT = naval support activity; NSY = naval shipyard; SFB = space force base.
[a] We include the wildfire case studies here for completeness but do not discuss this hazard in the rest of the report.

data through each asset's RPUID and can be uploaded into Hazus (a process discussed in Appendix A). Furthermore, buildings are likely at high risk of damage from hazards. Other high PRV asset categories—such as pavement, which includes runways and roads—could be temporarily disabled because of a hazard but are unlikely to be permanently damaged. Finally, resilience option costs for non-building assets are more challenging to estimate and require a more detailed study. The open literature, upon which we relied heavily to estimate resilience option costs, is skewed toward buildings. Table 2.3 summarizes our asset screening process.

Summary of Assets Included in Analysis

Table 2.4 displays the total number of buildings and total building PRV (in millions of U.S. dollars) for each installation included in the analysis. The results shown in the table account for the previously described screening and data preparation steps. The third column displays the total number of buildings included in the analysis, and the fourth column displays the total building PRV. The last column of the table displays the total PRV included in the analysis as a percentage of total installation PRV.

For most installations, the buildings included in the analysis constitute a large share of total installation PRV. However, for some installations, there are instances when our screening process filters out a majority of total installation PRV. There are several reasons why this is the case. In some cases, the relatively low percentage of covered installation PRV is because of the selection of asset types. Recall that our decision to focus on buildings was driven by data availability and computability with Hazus. However, at some installations, buildings make up a smaller share of total PRV. This is particularly true at NAVSUPPACT Norfolk NSY, where marine assets (e.g., wharves and piers) make up a significant portion of total installation PRV, which explains the relatively low PRV coverage in our analysis. If just counting building PRV at NAVSUPPACT Norfolk, our analysis covers 86 percent of assets.

In other cases, the low percentage of installation PRV is because of the nature of the hazard. Specifically, in the wildfire analysis, we focus on assets that have a non-zero burn probability per the USFS burn prob-

TABLE 2.3

Screening Process for Assets Within Installations

Included	Excluded	Reason for Exclusion
Buildings (68% of total PRV across installations) • Administrative and general buildings: 25% • Military housing: 19% • Storage and maintenance: 13% • Medical: 6% • Special purpose: 5%	• Port facilities (e.g., wharves): 5% of total PRV • Pavements (e.g., runways, roads): 5% of total PRV • Utilities (e.g., electrical substations, wastewater treatment plants): 10% of total PRV	• Unlikely to be destroyed; incompatible with Hazus • Incompatible with Hazus; unlikely to be destroyed • Incompatible with Hazus; missing PRV for most installations; poor GIS mapping quality

TABLE 2.4

Installation and Asset Data Summary (2022 U.S. dollars)

Cluster	Installation	Number of Buildings Included	Total PRV ($ millions)	Percentage of Total Installation PRV
Norfolk	JBLE	857	5,345.2	78%
	NWS Yorktown	600	1,596.4	76%
	NAVSTA Norfolk	982	6,896.7	60%
	NAVSUPPACT Hampton Roads	37	1,454.1	87%
	NAVSUPPACT Norfolk NSY	631	2,724.0	42%
Fort Bragg	Fort Bragg	5,304	18,962.1	72%

SOURCE: Authors' calculations using RPAD data provided by OSD CAPE on February 25, 2022.

NOTE: This table displays the total number of buildings, the total PRV, and the percent of installation PRV included in our analysis. These calculations are the result of the screening processes described in this report. JBLE = Joint Base Langley-Eustis; NAVSTA = Naval Station; NWS = Naval Weapon Station.

ability data. Most assets in the Colorado cluster have a burn probability of zero, and thus are not part of our analysis. We capture 100 percent of assets on burnable land in the wildfire analysis.

Some limitations of our screening process deserve additional discussion. First, we focus on PRV as a measure of building value, which does not include the value of building contents. We were unable to obtain building content value for all installations included in our analysis, although such data might be available and if so, could be used to augment our analysis. Some buildings could have a low PRV but high content value, and these types of buildings are filtered out of our analysis. Second, our screening process does not account for the mission value of an asset. Some assets could have low PRV but are extremely important to mission capabilities at an installation. Given our primary focus on the financial costs of disasters, we limit our treatment of mission implications to a qualitative discussion presented in Chapter 4. We discuss these and a few additional limitations in a section at the end of this chapter.

Resilience Options

We also screened resilience options. Many potential resilience options could limit exposure to—and damage from—hazards. For flood and wind damage, we focus on resilience options that can be assessed with Hazus; for wildfire burn probabilities, we focus on options for which reasonable cost estimates can be formulated. This screening process led to a variety of asset-level and area-wide resilience options. For coastal flooding and storm surge, we assess building-level floodproofing and an installation-wide floodwall.[15] For hurricane winds, we assess building-level roofing and hurricane shutters.[16] We discuss these in detail below.

Coastal Flooding and Storm Surge

For flooding and storm surge because of hurricanes, the resilience options we consider are constrained by what Hazus allows. However, even with the limitations of Hazus, we incorporate resilience options that could be implemented in a variety of ways. For instance, for flooding and storm surge, we investigate building-level resilience options and area-wide options. For hurricanes, we evaluate different types of building-level resilience options.

The flood-related resilience options we considered were 6 feet of building-level floodproofing and a 9-foot floodwall at the coastline. The analysis of the 6 feet of building-level flood protection involved re-coding each facility's first-floor elevation.[17] Elevating the first floor by 6 feet will mitigate flood damage at the same level as 6 feet of wet or dry floodproofing. The decision to use 6 feet for building-level flood protection assumed that the building structure would not be able to withstand hydrostatic pressure at flood levels greater than 6 feet.

We estimate the damage that might result with a 9-foot floodwall in place in a post-processing step, outside Hazus. Specifically, we use the flood height at the coastline to determine whether a 9-foot floodwall would protect the installation. If the flood height at the coastline is less than 9 feet, we assume that all facilities in the installation are protected from damage. On the other hand, if the flood height exceeds 9 feet, the floodwall provides no protection and the damage is identical to the baseline, no-mitigation results. Because the supply of water behind the floodwall is near infinite, it is reasonable to assume that if the flood overtops, the floodwall facilities will incur the same damage as if there were no floodwall. Additionally, the flood height at the coastline varies by location and depends on the event return period. Thus, a 9-foot floodwall could protect an installation in some situations but not in others.[18]

In total, we evaluate four flooding scenarios for each flood return period. Table 2.5 displays the flood and storm surge scenarios and their implementation in Hazus.

Hurricane Winds

We consider two resilience options for hurricanes: roof and window retrofits. To incorporate roofing retrofits, we re-coded each building's roof cover from the Hazus default (single-ply membrane) to a built-up roof. We also changed metal roof deck attachments from the Hazus default of standard to superior. Window retrofits involve adding hurricane shutters to each building. We also examined cases where both roof and window retrofits are included. Table 2.6 displays the resilience cases included in our analysis for hurricane winds.

---

[15] The building-level floodproofing included in our analysis is dry floodproofing that protects up to 6 feet of flooding. We use building-level floodproofing and dry floodproofing interchangeably in this report.

[16] In the Hazus technical manual, the specifications of the hurricane shutters are not based on the shutter type (e.g., roll-down, panel), but instead are described in terms of reduction in the average amount of damage to the building as a percentage. In our cost model of mitigation options, we assumed costs for roll-down hurricane shutters rated to withstand wind speeds of at least 110 mph (FEMA, "Hazus User & Technical Manuals," webpage, undated-b).

[17] Because of technical errors in the Hazus program, we were not able to directly model the 6-foot floodproofing option. Instead, we elevated each building by 6 feet to mimic the protection it would receive from 6 feet of floodproofing.

[18] Floodwalls and similar interventions can have negative environmental impacts (e.g., some concrete floodwalls can harm the surrounding ecosystem). Other nature-based alternatives might reduce such risks but were not considered in our analysis.

TABLE 2.5

Flood Resilience Options and Implementation in Hazus

Scenario	Implementation in Hazus
Baseline	Use Hazus defaults (building first-floor height is 1 foot)
Building-level, 6-foot dry floodproofing	Elevate building height by 6 feet, which has the same effect as using 6 feet of wet or dry floodproofing
Area-wide protection from 9-foot floodwall	Use default tides in Hazus and assume damage is same as baseline when floodwall is overtopped
Building-level, 6-foot protection and 9-foot floodwall	Use scenario-level tides and assume data are the same as 6-foot floodproofing if floodwall is overtopped

TABLE 2.6

Hurricane Wind Resilience Options and Implementation in Hazus

Scenario	Roof Type and Deck Attachment	Windows
Baseline	Use Hazus default	Use Hazus default
Retrofit roofing system only	Built-up roof with superior attachment	Use Hazus default
Installation of hurricane shutters only	Use Hazus default	Hurricane shutters
Retrofit roofing system and installation of hurricane shutters	Built-up roof with superior attachment	Hurricane shutters

Cost Model

Data

The data used to estimate resilience investment costs are based on two sources: (1) unit cost data from RSMeans, a construction cost-estimating software; and (2) unit cost data from natural hazard resilience literature. We drew on the literature to supplement data available in RSMeans to estimate costs for a given resilience option. For example, unit costs for the replacement of a roof and the installation of hurricane shutters are available in RSMeans, whereas unit costs for flood hazard mitigation, such as dry floodproofing or the installation of a floodwall, are not. Table 2.7 lists the data sources used to develop the cost estimates for the different resilience strategies.[19]

The unit cost data from both RSMeans and the literature are adjusted to 2022 dollars to account for inflation. Additionally, the unit cost data are adjusted to account for regional cost variations using city cost index (CCI) factors in RSMeans.[20] Further details on how the data sources were used are available in Appendix C of this report.

---

[19] We considered and reviewed other data sources pertaining to resilience options and hazard mitigation strategies in developing our cost model. OpenFEMA, for example, is a database that includes costs for completed projects funded through the FEMA Hazard Mitigation Grant Program. However, the data are reported as lump sum amounts per property, without any specifications on property size, which made it challenging to incorporate them into a cost model.

[20] The CCI is a resource incorporated into the RSMeans cost estimating tool that adjusts construction cost data to a specific location in the United States. To determine the specific CCI to use for each case study analysis, we identified the closest metropolitan city near the installation. For instance, the CCI location used for Fort Bragg is Durham, North Carolina.

TABLE 2.7

Cost Estimation Data Sources, by Hazard and Resilience Strategy

Hazard	Resilience Strategy	Data Source
Hurricane winds	Roofing replacement and hurricane shutter installation	RSMeans (Gordian, undated)
Coastal flooding and storm surge	Dry floodproofing	FEMA, 2007
Coastal flooding and storm surge	Floodwall	Aerts, 2018; FEMA, 2007

Methods

Our resilience option cost model includes unit costs derived from the data sources described above and from quantities of asset- and installation-specific site characteristics. The quantities for asset-specific characteristics (e.g., roof area, area of exterior windows) were collected using data from RPAD and by extrapolating with the RSMeans square foot estimator (SFE) tool.[21] The installation- or site-level quantities (e.g., length of coastline for a floodwall) were estimated using geospatial analysis. Table 2.8 lists the quantity inputs used in the model to estimate costs for a given resilience option.

To account for other costs, such as design and contractor fees, we added cost factors as percentages for general requirements, general conditions, design contingencies, general contractor markup, and other construction contingencies. For instance, an 11-percent factor was applied to roof replacement costs to account for general contractor markup fees (e.g., overhead, profit, and insurance). The choice of cost factors is based on the cost estimating format (CEF) tool used by FEMA. We annualized the final costs for each resilience strategy using a discount rate of 2.25 percent and a period based on the asset life of each resilience strategy according to our search of academic and industry literature. For instance, the asset life for a roof is assumed to be 30 years, whereas the asset life for dry floodproofing is assumed to be 22.5 years. Additional details on the methodology for the cost model can be found in Appendix C.

Limitations

Our approach has several limitations. First, focusing our analysis on buildings rather than other assets implies that we are not capturing the full costs of hazards or the full benefits of resilience options. For most installations, the buildings included in the analysis constitute a large share of total installation PRV. However, in some instances, our screening process filters out a large share of total installation PRV. This limitation is particularly relevant in the case of area-wide mitigation strategies, such as floodwalls, which would provide protection to non-building assets. Because our analysis does not account for averted damage from non-building assets, our estimates of the benefits of resilience might represent the lower bounds.

Second, we did not have access to data on building content value, which limits our ability to determine the benefits of resilience against hazards. As a result, building damage might not reflect the actual cost of a hazard. For instance, a low PRV building could house expensive equipment. Additionally, a low PRV building could be critical to mission readiness or other capabilities. We do not address how hazards or resilience options influence missions. In addition to the lack of data on building contents or mission importance, we also lack data on resilience options that an installation could have already implemented, which are needed to establish the true value of additional investments in resilience.

Third, our resilience cost estimates are not incremental over a baseline cost that includes ongoing maintenance and repair of existing facilities. Therefore, relative to ongoing maintenance and repair costs, the

---

[21] The SFE is a feature available in RSMeans to estimate the total cost of construction for a given building type and size.

TABLE 2.8

Asset and Site Inputs Used in Cost Model for Resilience Strategies

Hazard	Resilience Strategy	Asset or Site Input in Cost Model
Hurricane winds	Roof replacement	Roof area (SF) Roof perimeter (LF)
Hurricane winds	Installing hurricane shutters	Area of exterior windows (SF)
Coastal flooding and storm surge	Dry floodproofing	Building area (SF) Building perimeter (LF)
Coastal flooding and storm surge	Floodwall	Length of coastline within installation site boundary (km)

NOTE: LF = linear feet; SF = square feet.

resilience options we assess could be less expensive than they appear. That said, we also do not consider the ongoing maintenance costs of the resilience options. Instead, we consider only the annualized costs of implementing each option. There are likely annual ongoing costs associated with maintaining the resilience investment.

Fourth, because we were unable to find DoD cost factors in our analysis of resilience options, we leaned on commercial cost factors instead. We anecdotally compared our estimates of resilience option costs for one or two options with those provided by DoD subject-matter experts, but time constraints kept us from undertaking such a systematic comparison for each considered option.

Finally, the length of time over which a resilience option investment is repaid could significantly alter when and where that option yields cost savings. Without more information about how resilience implementation is financed, it is difficult to account for the relevant repayment period. For example, annual payments will be low if the cost of implementing a resilience option can be spread over the option's lifetime. These low annual payments compare favorably to the expected yearly averted damage. However, if the entire cost of implementing the resilience option must be paid immediately and cannot be spread out over time, the comparison with expected annual averted damage will be much less favorable. We address this limitation by first assuming that implementation costs can be spread evenly across the entire lifetime of the option. This assumption produces the most favorable conditions for the resilience option compared with the expected annual averted damage. Then, for the resilience options that yield cost savings in this favorable scenario, we assess whether the option would still yield cost savings if all costs were paid in the first year rather than spread over the option's lifetime.

CHAPTER 3

Illustrative Results

We present illustrative results of the analysis in several stages, summarized in Figure 3.1. First, we discuss the building damage cost estimates without additional resilience options in place. We refer to these estimates as the *baseline damage*. Then, we present the damage estimates with resilience options in place. We refer to the difference between baseline damage and damage with a resilience option as the averted damage of the resilience option. Next, we present cost estimates for each resilience option's implementation. Finally, we compare each resilience option's averted damage with its implementation costs, assessing how much more frequently hazards need to occur for the option to yield cost savings.

FIGURE 3.1
Outline of Results Chapter

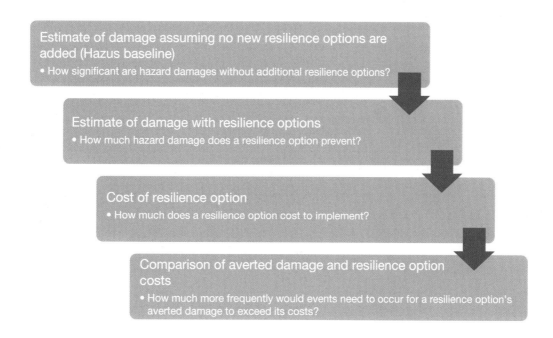

Damage Costs With and Without Resilience Options in Place

Tables 3.1 and 3.2 show total installation damage (in millions of U.S. dollars) for events related to floods and hurricanes, respectively. For each hazard, the tables present damage estimates for the baseline scenario (without resilience options) and for each resilience option scenario.

Coastal Flood and Storm Surge

The first half of Table 3.1 (columns two through five) shows that baseline coastal flood damage increases quickly between 10- and 50-year events, particularly for NAVSTA Norfolk and JBLE. A 500-year coastal flood is estimated to cause more than $400 million in damage at both NAVSTA Norfolk and JLBE. This is equivalent to approximately 8 percent of total PRV for JBLE and 6 percent of total included PRV at NAVSTA Norfolk. However, NWS Yorktown is more protected from coastal floods: A 500-year event at this installation is estimated to cause $170,000 in damage (0.01 percent of total included PRV).

The effectiveness of each resilience option also varies by installation and return period event. For instance, 6 feet of building-level floodproofing reduces damage costs significantly across nearly all installations. Building-level floodproofing is relatively less effective at NWS Yorktown, but the baseline damage is also quite small at this installation. At all installations, a 9-foot floodwall at the coast would protect all buildings up to a 50-year coastal flood event. However, the floodwall would break for 100- and 500-year events, resulting in the same damage costs as the baseline in those cases. When a 9-foot floodwall and 6 feet of building-level floodproofing are used together, all installations are protected from more minor coastal floods and withstand more-limited damage for more frequent events.

The second half of Table 3.1 (columns six through eight) displays the results of the storm surge analysis. Here, the annual return periods are shown as a range because they are based on hurricane categories, which occur at different frequencies across locations. For instance, a Category 1 hurricane storm surge event has a 75-year return period at NWS Yorktown, but it has a 25-year return period at NAVSUPPACT Norfolk. See the "Step 3c: Storm Surge" section in Appendix A for a description of how we calculated storm surge return periods. As with coastal floods, damage costs from storm surges are high at NAVSTA Norfolk and JBLE. However, most installations face larger damage costs from storm surge events than from coastal flood events. Coastal floods and storm surge are disparate hazards that can occur in differing flood extents, both in terms of area exposed and depth of flooding. As a result, an installation could be relatively protected from coastal floods but highly exposed to storm surges depending on the natural geographic characteristics of the area.

For storm surge events, a 9-foot floodwall at the coast is relatively effective at reducing damage costs for minor events but is generally less effective for the 150–500- and 1,000-plus-year events. In some cases, the 9-foot floodwall option provides no protection from storm surge damage. For example, at NWS Yorktown, damage with and without a 9-foot floodwall is identical because the floodwall provides no protection for the installation. Similarly, at NWS Yorktown, the 6-foot floodproofing option provides very little protection. However, the 6-foot floodproofing option provides more protection at other installations. For instance, at JBLE, 6 feet of floodproofing on all buildings reduces the damage across storm surge return periods by approximately 75 percent, on average.

For both coastal floods and storm surge events, we did not assess the implications of a 9-foot floodwall at NAVSUPPACT Hampton Roads because there are no coastlines around that installation.

TABLE 3.1

Flood Event Building Damage Across Installations and Return Periods

Installation Scenario	Coastal Flooding Annual Return Periods (2022 $ millions)				Storm Surge Annual Return Periods (2022 $ millions)		
	10 Years	50 Years	100 Years	500 Years	25–75 Years	150–500 Years	1,000+ Years
NAVSTA Norfolk							
Baseline	2.42	182.39	255.14	402.86	0.00	146.44	576.55
6-ft building-level flood proofing	1.14	1.60	1.80	3.80	0.00	34.10	115.70
9-ft floodwall	0.00	0.00	255.14	402.86	0.00	146.44	576.55
Floodwall and building-level floodproofing	0.00	0.00	1.80	3.80	0.00	34.10	115.70
NWS Yorktown							
Baseline	0.07	0.11	0.13	0.17	0.00	22.41	22.74
6-ft building-level floodproofing	0.03	0.05	0.06	0.08	0.00	20.11	22.32
9-ft floodwall	0.00	0.00	0.13	0.17	0.00	22.41	22.74
Floodwall and building-level floodproofing	0.00	0.00	0.06	0.08	0.00	20.11	22.32
NAVSUPPACT Norfolk							
Baseline	6.91	28.18	95.85	282.61	101.60	116.25	266.80
6-ft building-level floodproofing	2.90	4.13	4.58	7.26	0.35	46.54	103.61
9-ft floodwall	0.00	0.00	95.85	282.61	0.00	0.00	266.80
Floodwall and building-level floodproofing	0.00	0.00	4.58	7.26	0.00	0.00	103.61
JBLE							
Baseline	2.61	65.33	165.90	422.39	170.43	438.30	819.15
6-ft building-level floodproofing	0.00	0.03	0.99	3.48	9.83	75.10	409.72
9-ft floodwall	0.00	0.00	165.90	422.39	0.00	438.30	819.15
Floodwall and building-level floodproofing	0.00	0.00	0.99	3.48	0.00	75.10	409.72
NAVSUPPACT Hampton Roads							
Baseline	0.00	0.05	2.91	15.13	359.10	525.28	703.73
6-ft building-level floodproofing	0.00	0.00	0.00	0.00	0.36	100.70	499.30

Table 3.1—Continued

Installation Scenario	Coastal Flooding Annual Return Periods (2022 $ millions)				Storm Surge Annual Return Periods (2022 $ millions)		
	10 Years	50 Years	100 Years	500 Years	25–75 Years	150–500 Years	1,000+ Years
9-ft floodwall[a]							
Floodwall and building-level floodproofing[a]							

SOURCE: Authors' calculations using RPAD and Hazus data.

NOTE: This table displays the results from the coastal flooding and storm surge scenarios. Data from individual buildings is aggregated to the installation-resilience option-return period level.

[a] Not assessed because there are no coastlines around NAVSUPPACT Hampton Roads.

Hurricane Winds

Table 3.2 displays the building damage costs for hurricane winds, in millions of U.S. dollars, for the baseline and resilience option scenarios. The amount of wind damage at Fort Bragg is highest across all return period events, largely because of Fort Bragg's size and high total PRV. However, damage per building (total damage per number of buildings) is also highest at Fort Bragg. On the other end of the spectrum, NWS Yorktown is relatively protected from hurricane damage. A 1,000-year event at NWS Yorktown produces only $99 million in damage, while a similar event at other installations produces more than $350 million, and sometimes billions of dollars, in damage.

Across installations and hurricane return periods, the resilience options provide some protection from hurricane wind damage, but no option eliminates damage. The most effective option includes retrofitting the roofing and windows on all buildings. For instance, retrofitting both the roofing and windows on all buildings at Fort Bragg reduces damage by an average of 61 percent across hurricane wind return periods. The same package of options has a smaller effect on damage reduction at other installations. The average amount of baseline damage reduction from retrofitting roofing and window systems at installations in the Norfolk cluster ranges from 19 percent to 23 percent.

Resilience Option Costs

Tables 3.3 through 3.6 summarize the total and annualized costs in millions of (2022) dollars for the resilience options for each installation.[1] The installations with the highest and lowest resilience option costs are Fort Bragg and NAVSUPPACT Hampton Roads, respectively. The number, size, and type of the assets were primary factors in the high costs for hurricane resilience options (roof replacement) and for building-level floodproofing. Recall that Fort Bragg has the highest number of assets (5,304) and NAVSUPPACT Hampton Roads has the lowest number (37 buildings). Costs for floodwalls are highest for JBLE and lowest for NAVSUPPACT Hampton Roads and Fort Bragg. JBLE accounts for two installation sites that are both adjacent to a coastline, whereas the site boundaries for Fort Bragg and NAVSUPPACT Hampton Roads are located more inland and away from the coastline.

---

[1] We use a discount rate of 2.25 percent. See the end of this chapter and Appendix C for details regarding our choice of time frame to annualize option costs.

TABLE 3.2

Hurricane Wind Event Building Damage Costs Across Installations and Return Periods

Installation	Hurricane Wind Annual Return Periods (2022 $ millions)			
	10 Years	100 Years	500 Years	1,000 Years
NAVSTA Norfolk				
Baseline	5.26	328.07	538.84	1,411.35
Retrofitting roofing	4.49	208.39	381.82	1,178.41
Installing hurricane shutters	5.05	293.60	446.37	935.52
Retrofitting roofing and installing hurricane shutters	4.31	173.26	282.30	638.06
NWS Yorktown				
Baseline	0.03	4.47	160.20	99.13
Retrofitting roofing	0.03	3.02	127.80	76.57
Installing hurricane shutters	0.03	3.92	92.54	62.30
Retrofitting roofing and installing hurricane shutters	0.03	2.48	54.98	37.09
NAVSUPPACT Norfolk				
Baseline	2.07	79.04	139.43	638.13
Retrofitting roofing	1.90	53.13	97.90	520.15
Installing hurricane shutters	1.85	62.08	97.53	339.72
Retrofitting roofing and installing hurricane shutters	1.70	35.16	54.50	189.22
JBLE				
Baseline	1.56	40.82	266.16	514.12
Retrofitting roofing	1.49	27.35	189.70	391.87
Installing hurricane shutters	1.54	33.44	170.56	304.69
Retrofitting roofing and installing hurricane shutters	1.43	19.58	85.89	157.37
NAVSUPPACT Hampton Roads				
Baseline	1.10	60.69	96.12	362.66
Retrofitting roofing	0.92	35.74	61.12	311.90
Installing hurricane shutters	1.07	56.41	85.80	275.05
Retrofitting roofing and installing hurricane shutters	0.91	30.95	49.42	192.89

Table 3.2—Continued

Installation	Hurricane Wind Annual Return Periods (2022 $ millions)			
	10 Years	100 Years	500 Years	1,000 Years
Fort Bragg				
Baseline	6.63	351.08	231.24	3,525.09
Retrofitting roofing	6.59	91.42	177.81	3,059.46
Installing hurricane shutters	6.40	204.60	148.49	1,613.84
Retrofitting roofing and installing hurricane shutters	6.22	54.65	99.11	801.15

SOURCE: Authors' calculations using RPAD and Hazus data.
NOTE: This table displays the results from the hurricane wind scenarios. Data from individual buildings is aggregated to the installation-resilience option-return period level.

TABLE 3.3

Total and Annualized Costs for Hurricane Resilience (Roof Replacement) for Buildings

Installation	Total ($ millions)	Annualized Costs ($ millions)
Fort Bragg	733.70	33.90
JBLE	225.60	10.40
NAVSTA Norfolk	249.70	11.50
NAVSUPPACT Hampton Roads	13.40	0.62
NAVSUPPACT Norfolk NSY	150.30	6.90
NWS Yorktown	108.60	5.00

TABLE 3.4

Total and Annualized Costs for Hurricane Resilience (Hurricane Shutter Installation) for Buildings

Installation	Total ($ millions)	Annualized Costs ($ millions)
Fort Bragg	405.90	18.80
JBLE	69.70	3.20
NAVSTA Norfolk	63.90	2.90
NAVSUPPACT Hampton Roads	8.80	0.41
NAVSUPPACT Norfolk NSY	30.60	1.40
NWS Yorktown	19.90	0.92

TABLE 3.5

Total and Annualized Costs for Coastal Flooding and Storm Surge (Dry Floodproofing) for Buildings

Installation	Total ($ millions)	Annualized Costs ($ millions)
Fort Bragg	264.20	15.10
JBLE	55.20	3.20
NAVSTA Norfolk	54.10	3.10
NAVSUPPACT Hampton Roads	2.96	0.17
NAVSUPPACT Norfolk NSY	32.40	1.90
NWS Yorktown	28.50	1.60

TABLE 3.6

Total and Annualized Costs for Coastal Flooding and Storm Surge (Floodwall)

Installation	Total ($ millions)	Annualized Costs ($ millions)
Fort Bragg	0.00	0.00
JBLE	249.80	8.40
NAVSTA Norfolk	89.10	3.00
NAVSUPPACT Hampton Roads	0.00	0.00
NAVSUPPACT Norfolk NSY	19.70	0.66
NWS Yorktown	45.70	1.50

Comparing Averted Damage and Resilience Option Costs

Next, we return to Equation 1, shown in the overview of Chapter 2, to assess the damage costs that a resilience option averts over its lifetime, which we compare with the resilience option's annualized cost. We start by assuming that hazard return periods remain at their current level; then we assess how the averted damage calculations change when hazards become more frequent. This allows us to identify how much more frequently a hazard would need to occur for a resilience option to have positive annual savings (i.e., when the averted lifetime damage is greater than the cost of implementation).

Coastal Flooding Savings

To build intuition, we present the annual savings from resilience (annualized averted damage minus annualized cost of implementing an option) for coastal flooding resilience options first by assuming historical flood frequencies, and then by assuming that flooding events occur five times more frequently (Figure 3.2). The y-axis shows the difference between the damage the resilience option averts over its lifetime and the one-time cost of implementation. Naturally, resilience options become more cost-effective when flood events occur more frequently. However, even with historical flood frequency levels, some resilience options appear promising. For example, both the 9-foot floodwall and the 6 feet of building floodproofing create positive annual

FIGURE 3.2

Annual Savings Calculations for Coastal Flooding Options with Different Probabilities

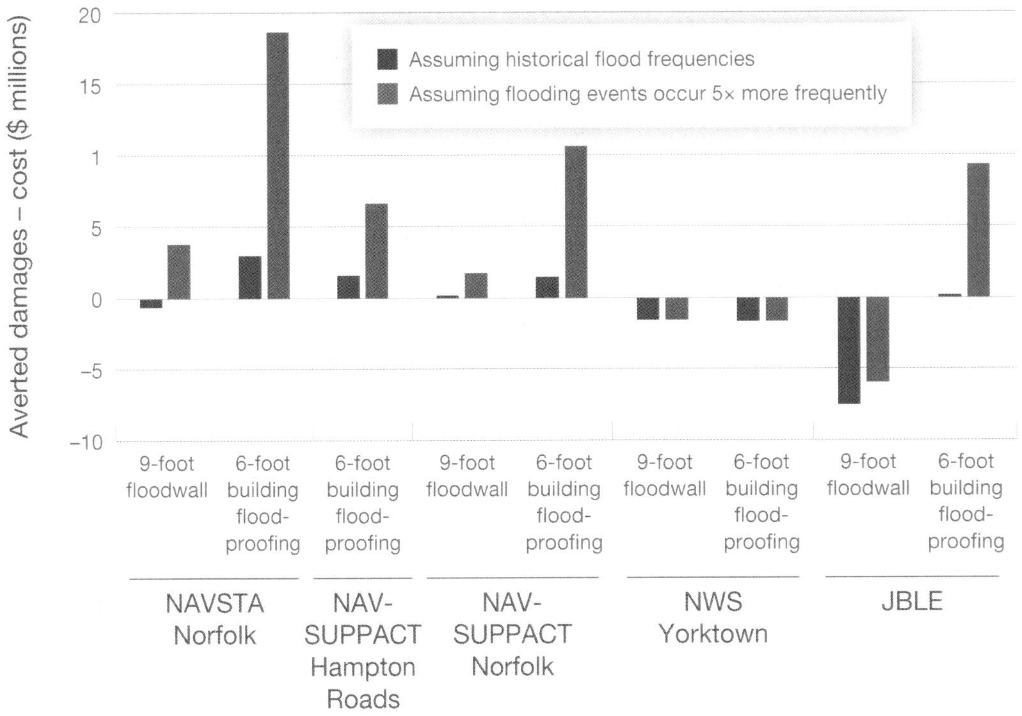

NOTE: This figure displays the annual savings calculations for coastal flooding resilience options under current and historical flood frequencies, and for when flood frequencies increase by a factor of five.

savings at NAVSUPPACT Norfolk. As floods become more frequent and intense, these options become even more promising. However, even if floods become five times more frequent, neither resilience option yields positive savings at NWS Yorktown.

Next, we calculate how much more frequently an event of a given severity would need to occur for a given resilience option to yield cost savings over the option's lifetime at an installation. The results are shown in Table 3.7. The values in the table display the frequency multiplier required for a resilience option to produce a positive annual savings. A 1x in a cell indicates that the resilience option (column) yields cost savings at a given installation (row) if the frequency with which hazard events (e.g., coastal floods) of a given severity continues to be the same as the historical record (i.e., the frequency multiplier is 1). However, 6 feet of building floodproofing at NWS Yorktown becomes cost effective only if coastal floods at the installation become 18 times more frequent (i.e., the frequency multiplier is 18). The table also highlights resilience options that are never cost effective. For example, a 9-foot floodwall will never result in positive annual savings for storm surge events at NAVSTA Norfolk, regardless of how much more frequent hurricane storm surge events become. This is because a 9-foot floodwall provides no additional protection from storm surge events (i.e., above what is provided naturally by the region's topography), regardless of return period, at NAVSTA Norfolk. This situation is similar at NWS Yorktown, where a 9-foot floodwall never has positive annual savings because it provides no additional protection.

TABLE 3.7

Increase in Hazard Frequency Needed for Resilience Option Cost Savings, Assuming That Resilience Option Costs Are Evenly Spread over the Option's Lifetime

Installation	Coastal Flooding			Hurricane Winds			Storm Surge		
	6-ft Building Floodproof	9-ft Floodwall	Both	Roof Upgrade	Window Shutters	Both	6-ft Building Floodproof	9-ft Floodwall	Both
NAVSTA Norfolk	1x	2x	1x	4x	4x	4x	5x	Never	9x
NAVSUPPACT Hampton Roads	1x			2x	3x	2x	1x		
NAVSUPPACT Norfolk	1x	1x	1x	11x	3x	9x	1x	2x	2x
NWS Yorktown	18x	13x	18x	15x	7x	15x	76x	Never	82x
JBLE	1x	9x	5x	14x	9x	13x	1x	6x	3x
Fort Bragg				13x	6x	11x			

SOURCE: Authors' calculations using RPAD and Hazus data.

NOTE: Blank cells mean that the installation does not experience the hazard because of its geographic location.

For coastal flooding events, 6 feet of building-level floodproofing shows promise as a resilience option. Apart from NWS Yorktown, building-level floodproofing creates positive annual savings with current flood event probabilities. At NAVSUPPACT Norfolk, all resilience options create positive annual savings with current flood event probabilities. The 9-foot floodwall option pays off only if coastal floods increase in frequency, although, in some cases, the frequency multiplier is small.

Storm Surge Savings

The resilience options for storm surge events are less promising than for coastal flooding. Because these two types of flood events have different return periods and different inundation levels, the finding that resilience options have different savings-yielding frequency multipliers is not surprising. However, we find that the 6-foot building floodproofing option yields cost savings under current storm surge event probabilities at NAVSUPPACT Hampton Roads, NAVSUPPACT Norfolk, and JBLE. The 9-foot floodwall offers no protection at NAVSTA Norfolk or at NWS Yorktown and, thus, never yields cost savings at these installations.

Hurricane Winds

For hurricane winds, we find that no resilience option provides positive annual savings until hurricane wind events become at least twice as frequent as they are currently. Across installations, window upgrades appear to pay off sooner than do roof upgrades. Similarly, across installations, window upgrades are less costly to implement than roof upgrades, making the savings-yielding frequency multiplier smaller for window upgrades, even though roof upgrades offer more protection from extreme hurricane winds.[2]

---

[2] The estimated roof replacement costs include the costs of replacing the roofing insulation layer, replacing the roof covering layer, and replacing the roof flashing and other seams. We consider this estimate a lower bound because it does not include the costs of replacing the roofing deck, which is typically not required unless the roof is in poor condition. As a sensitivity exercise, we assess how the inclusion of roofing deck replacement costs influence the results and find that it has a relatively minor

The Effect of the Repayment Period

As previously discussed, our calculations of whether a resilience option yields cost savings depends on how the implementation costs are financed. The results in Table 3.7 assume that the costs of implementing an option can be evenly spread over the option's lifetime, resulting in the lowest possible annual payments. However, if the cost of implementing the resilience option must be paid off sooner, leading to higher annual payments, the frequency multipliers shown in Table 3.7. will change.

To assess how much the repayment period influences the results, we calculate the savings-yielding frequency multipliers for the most promising options shown in Table 3.8—6 feet of building floodproofing at NAVSUPPACT Hampton Roads, NAVSUPPACT Norfolk, and JBLE. These resilience options yield cost savings under current event probabilities for both coastal floods and storm surge events. Other resilience options do not protect installations from multiple hazards. Additionally, other resilience options do not yield cost savings, even with the most favorable implementation cost repayment plan.

In Table 3.8, we show the savings-yielding frequency multipliers for the select group of installations. We show two alternative payment plan scenarios. First, we assume that the cost of implementing the 6-foot building floodproofing must be paid over 12 years (half of the option's lifetime). Then, we assume that the cost of implementation must be fully paid in the first year. Unsurprisingly, the frequency multipliers increase in most cases. For instance, if the cost must be paid in the first year, coastal floods must become at least twice as frequent for 6 feet of building floodproofing to yield savings at any installation. However, we find that 6 feet of building floodproofing yields savings under current event probabilities for storm surge at NAVSUPPACT Hampton Roads, even if the implementation cost must be paid in the first year. This is largely attributed to NAVSUPPACT Hampton Roads' relatively low implementation cost (shown in Table 3.5), as well as the significant amount of avoided damage the resilience option offers the installation against storm surge events (shown in Table 3.1).

As also shown in Table 3.1, 6 feet of building floodproofing fully protects NAVSUPPACT Hampton Roads from coastal flood damage. However, the installation is less exposed to coastal flood damage than it is to storm surge damage. In other words, the avoided damage (damage without resilience minus damage with resilience) is smaller for coastal floods than it is for storm surge. As a result, coastal floods would need to become at least twice as frequent for 6 feet of building floodproofing to yield cost savings at NAVSUPPACT Hampton Roads. This finding highlights the importance of taking a cross-hazard view when evaluating resilience options that are effective against multiple hazards.

TABLE 3.8

Increase in Hazard Frequency Needed for 6-Foot Building Floodproofing Cost Savings, Assuming Option Costs Are Paid over Some Portion of the Option's Lifetime

	Coastal Flooding		Storm Surge	
Repayment period	12 years	1 year	12 years	1 year
NAVSUPPACT Hampton Roads	1x	2x	1x	1x
NAVSUPPACT Norfolk	2x	11x	4x	11x
JBLE	2x	13x	2x	11x

SOURCE: Authors' calculations using RPAD and Hazus data.

effect: Most frequency multipliers increase by one or two. The one exception is NAVSTA Norfolk, which sees its frequency multiplier increase from 4 times to 12 times.

Limitations on Applying the Results

Although our analysis identifies some promising investments, additional considerations should drive decisionmaking. Our analysis was necessarily abstracted from comprehensive installation realities (to be tractable and transparent). Therefore, the relative attractiveness of particular resilience options at particular installations, as revealed by our analysis, might increase or decrease with the consideration of additional factors not included in the analysis.

Such factors include

- **Current level of resilience at installations.** Although our interviews with installation subject-matter experts provided anecdotal evidence of resilience measures having been taken, we were unable to find any central, authoritative repository of such information. The implication of not considering this baseline level of resilience is that we are likely to have overestimated the size of the resilience investment required at each location to cope with the relevant hazard(s).
- **Mission implications and other installation goals.** In Chapter 4, we include a qualitative discussion of mission implications and other goals, such as environmental stewardship or Service member quality of life, that might change the decision calculus of whether and where to invest in installation resilience. For example, an installation might choose to invest only in resilience options in a limited set of high-cost, highly valued, and mission-critical buildings, and include the mission cost savings (e.g., costs of having to relocate the mission while a building is being repaired) and other benefits in a cost-benefit analysis of the options. Doing so would lead to different results about the value of such investments.
- **Budget, financing, and engineering constraints.** Such constraints ultimately govern whether and when resilience projects are executed and cannot be assessed at a high level without specific, localized inputs.
- **Damage to non-building assets (e.g., pavements, utilities) and to building contents.** We noted in Chapter 2 that we did not consider damage to the full set of assets at every installation for several reasons. The implication of this omission is that our results likely underestimate the benefits afforded by the resilience options.
- **Maintenance activities and cycles.** Resilience upgrades could be folded into existing maintenance cycles. For example, the next time a roof is ready for replacement, it could be upgraded to withstand higher wind speeds—and at an installation where high winds are a known issue, it likely would be upgraded. Aligning resilience-focused upgrades with existing processes is likely to reduce some overhead costs and out-of-cycle personnel costs that might otherwise be incurred.
- **Age or remaining life span of assets.** We do not account for these factors when deciding which assets receive resilience options. One likely implication of not making these distinctions is that we are overestimating the costs of the resilience options, because it might not make sense to include all assets in our tally of the assets that should be upgraded. For instance, it probably does not make sense to upgrade a facility that is on the verge of collapse because of age. A more prudent investment (mission impact notwithstanding) might be to mothball or demolish it.

In summary, the results presented in Tables 3.7 and 3.8 should be viewed as an early step in identifying installations and resilience options that are good candidates for deeper analysis. In Chapter 4, we discuss one of the previously mentioned factors—mission implications.

CHAPTER 4

Implications for Missions and Other Installation Goals

DoD's mission is to prepare the armed forces to fight the nation's wars, and everything it does, at least in theory, is oriented toward that mission. Yet DoD as a whole, and each of its installations, must attend to and balance other goals. Often, the most prominent of these goals, after mission, is cost-effectiveness. When DoD acts in fiscally mindful ways, it typically does so to preserve resources to direct to other missions and activities. Additional installation goals include environmental compliance and stewardship, as well as quality of life for Service members and their families.

In the preceding chapters, we have focused squarely on the financial implications of disasters and disaster preparedness for installations, which was the primary focus of our analysis.

In this chapter, we orient ourselves toward these other installation goals. We primarily focus on the mission impacts of extreme weather events, considering how mission impact could direct DoD's decision calculus in the same way that financial logic has in our analysis thus far, or in other directions. We also describe how other goals trade off against mission and cost.

Disruption to Operations or Support Activities

In the Preparation Phase

Many events come with days, if not weeks, of warning. Hurricanes (bringing high winds and storm surge) are usually preceded by days of warning, albeit often with significant uncertainty about the precise location of highest impacts might be experienced. Heavy rains might or might not come with sufficient warning. Other hazards might only present hours of notice but recur seasonally, such as wildfires or tornadoes. Preparations can be made each season, and resources can be stockpiled (e.g., sandbags for flooding, boards for windows) and deployed even with only a few available hours of preparation. Some of these measures will require an urgency that disrupts operations. Steps could be taken to ready equipment and facilities that directly support missions (e.g., securing and protecting equipment and facilities) or could involve diverting mission and support personnel to implement measures elsewhere on the base (e.g., in their own homes to ensure that they can safely get through the storm). Although preparing for an uncertain event would somewhat disrupt the mission, these disruptions would likely pale in comparison with what would be experienced without any preparations were the event to occur.

During the Event

Events themselves can last hours or even days, and for most DoD missions and activities, which involve sizable weapon systems, equipment, and vehicles that are extremely vulnerable to weather events, operations must cease to protect personnel and equipment.[1] Even if an event does not cause the kind of acute damage experienced from hurricanes, tornadoes, or similar occurrences, conditions such as extremely high temperatures can make training activities impractical or impossible.[2] The extent of mission disruption during an event depends on the ability of the mission itself to be moved to, or replicated at, a geographically separate location.[3] Some missions have entirely redundant locations (e.g., a subset of space operations); others have a distributed, networked design that can shift workloads to some extent (e.g., the Air Force Distributed Common Ground System); and still others can move their mission equipment to operate from alternative locations, albeit with some degradation or inefficiency (e.g., aircraft training).[4]

Post-Event Recovery

The extent of, and timeline for, post-event recovery can vary widely and depends greatly on human interventions and the severity of the event. Recovery can be brief—hours or days—if little damage has occurred. Recovery can also last months or even years, as evidenced by recent high-profile events, such as Hurricane Sally in 2020, from which the recoveries are still underway.[5] In some cases, the personnel at a given installation might assist with recovery activities, either for their own personal homes and communities or for the installation where they operate. Some recovery entails clearing debris from the base, which can make roads, pathways, and other areas inaccessible or dangerous.[6] The safety of base personnel is already a priority, and much of the equipment the military operates is quite sensitive and must only be operated under particular conditions.

Damage to, or Inaccessibility of, Equipment, Personnel, Facilities, or Supporting Equipment

One of the most obvious ways disasters can affect missions is by damaging facilities, or the equipment and supplies contained therein, that serve and support the mission, thus rendering them unusable until repaired or replaced. Hurricane Michael, which struck Tyndall AFB in 2018, damaged every facility on base, stripping equipment and supplies from buildings and scattering them far and wide, as well as damaging critical

---

[1] DoD, *Department of Defense Climate Adaptation Plan*, September 1, 2021.

[2] Shirley V. Scott and Shahedul Khan, "The Implications of Climate Change for the Military and for Conflict Prevention, Including Through Peace Missions," *Air and Space Power Journal—Africa and Francophonie*, 3rd Quarter, 2016.

[3] See Anu Narayanan, Michael J. Lostumbo, Kristin Van Abel, Michael T. Wilson, Anna Jean Wirth, and Rahim Ali, *Grounded: An Enterprise-Wide Look at Department of the Air Force Installation Exposure to Natural Hazards: Implications for Infrastructure Investment Decisionmaking and Continuity of Operations Planning*, RAND Corporation, RR-A523-1, 2021, pp. 79–94.

[4] Schriever Space Force Base, "3rd SOPS Backup Center Ready to Take On All Ops," undated; U.S. Air Force, "Air Force Distributed Common Ground System," October 2015; Geoff Ziezulewicz, "Hurricane Ian Prompts Evacuation of Navy Ships and Aircraft," *Navy Times*, September 27, 2022; Tara Copp, "71 of the Navy's T-45 Trainer Jets Could Not Evacuate Storm," *Navy Times*, August 25, 2017; Tyler Rogoway, "Setting the Record Straight on Why Fighter Jets Can't All Simply Fly Away to Escape Storms," *War Zone*, December 1, 2019.

[5] Naval Facilities Engineering Systems Command, "NAVFAC Southeast Stands Up Resident Officer in Charge of Construction Sally for Hurricane Repair in Pensacola," September 16, 2022; Greg Hadley, "Tyndall and Offutt Rebuilds Need More Funds; Projected End Date in 2027–'28, *Air Force Magazine*, May 2, 2022.

[6] Caitlin M. Kenney, "Air Force, Marine Corps Warn of Critical Readiness Concerns Without Supplemental Disaster Funding," *Stars and Stripes*, April 30, 2019; Jan Wesner Childs, "Rebuilding of Tyndall, Offutt Air Force Bases After Storms Stalled Due to Lack of Funds," Weather Channel, May 1, 2019.

mission equipment, such as aircraft, that were not evacuated in time.[7] Even after the initial chaos subsided and initial recovery operations cleared the worst debris, the overall losses and damage prevented, or at least degraded, operations until key facilities and equipment were repaired or replaced. Recovery efforts could be prioritized to focus on the most mission-critical needs first, but those efforts could take months or years for the most extreme events.

Even if critical facilities and operational systems are preserved, mission disruption and degradation can still occur if those facilities and systems are rendered inaccessible by road or taxiway.[8] This can occur from debris strewn about by high winds, for example; most of the worst debris can be cleared away in days or weeks in extreme cases. Potentially more consequential is flooding of access roads and damage to the road system. Although less likely, roads could be damaged in a flooding scenario. More likely could be the erosion and degradation of culverts, rendering parts of roads impassable, and damage to bridges. In the Offutt AFB flood, the runway was flooded, as were roadways. At Langley AFB, flooding occurs annually on the portions of the base that are not protected by a powerful pumping station.[9] RAND researchers analyzed flooding scenarios at several Air Force and Space Force bases as part of previous research for the Department of the Air Force.[10] That research showed some cases where critical operations or support facilities would be affected by flooding, but in other cases, flooding only affected access roads, leaving more-critical facilities unaffected. Thus, protecting what would conventionally be considered the most mission-critical facilities might still leave the mission itself vulnerable to disruption.

Implications of Favoring Mission Impact Versus Financial Savings as the Primary Driver of Resilience Investment Prioritization

How would prioritizing investments for mission impact differ from an approach that favors financial savings?

First, mission-driven investments would focus, generally, on mission-critical infrastructure, with less emphasis on either the value of that infrastructure or the cost of implementing resilience options. A purely financial trade-off, such as the one we have explored in this report, aims to economically justify any money invested against the expected cost savings. A truly mission-critical infrastructure asset or system could justify investment in a resilience option, even when the cost of the resilience option rivals the replacement cost of the asset or system. Take, for instance, a communication pathway, energy source, or operations center. Establishing a system that is entirely redundant to one of these tasks essentially doubles the cost of providing that function under normal conditions. It is only when damage (or the threat thereof) occurs that the redundant asset comes into play.

Second, mission-driven investments are likely to consider disruptions that cause little to no damage. As with the example of roads and other access ways, flooding can cause severe disruption by blocking paths to mission-critical infrastructure, even if no damage is ultimately caused. In some cases, flooding might be confined to specific areas, to the extent that the original road and access planning did not account for risks from potential flooding. With increased flooding, pathways that are normally accessible under typical conditions become inaccessible. This could incentivize building alternative roadways that can access critical areas via other paths, such as raising extremely vulnerable pathways to avoid flooding or even relocating critical

---

[7] Becky Sullivan, Noah Caldwell, and Ari Shapiro, "Nearly 8 Months After Hurricane Michael, Florida Panhandle Feels Left Behind," National Public Radio, May 31, 2019.

[8] Narayanan et al., 2021, pp. 42–49.

[9] Langley AFB personnel, interviews with the authors, April 28, 2022.

[10] Narayanan et al., 2021.

facilities to essentially be invulnerable to flooding. This example serves as another case where the financial value of the vulnerable asset (in this case roads) is not really at risk, but mission owners might be willing to take expensive measures to ensure access to mission-critical infrastructure. Additionally, such logic could incentivize investments in recovery speed, which brings us to our third observation.

Mission-driven investments would also focus on recovery speed to minimize the duration of a disruption in cases where disruptions were inevitable. Minimizing recovery speed can have varying benefits for different types of assets. Take flooding for instance. For many vertical structures, the duration of exposure to floodwater and flow are key factors in determining the extent of damage.[11] The longer assets and their contents remain in contact with water (especially saltwater, which would result from hurricane storm surges), the more damage that is caused, and the more expensive it is to repair the damage. A resilience option, such as a water pump, that could reduce the exposure of vertical structures to floodwaters (both in terms of quantity and duration) would reduce damage to structures and the contents within. In this case, a pure comparison of the costs of investing in the resilience option versus the damage costs that result from a disaster might serve as a sufficient basis for deciding whether to proceed with the resilience investment. This is precisely the logic that drove us to focus on vertical structures in a financially focused analysis.

Conversely, for horizontal infrastructure, such as access ways and runways, the asset might experience little damage in relation to the extent of mission disruption. Features of the water pump—its ability to rapidly empty an area of floodwater and return to service areas such as roadways, taxiways, runways, parking aprons, and other areas that are vulnerable to flooding—make it an attractive candidate for investment. The resilience benefit would take the form of quicker access to mission-critical assets, paths which could have otherwise remained blocked long enough to significantly disrupt operations. In this case, although such a resilience option is likely to be relatively cheap compared with retrofitting entire buildings, this cost advantage might not be what drives investment. Instead, what makes this resilience option attractive is the promise of improving mission assurance in the face of disaster.

Although deeper examination of the mission implications of investments in resilience (or lack thereof) is out of scope for our analysis, this aspect merits serious consideration when decisions are made regarding where and how much to invest in installation resilience to climate change.

Other Goals for Installations

DoD and the Services have other goals that are relevant to climate change and the related adaptation requirements that are placed on installations. All of the Services have climate adaptation strategies and plans, each with goals and lines of effort that include, but are not limited to, installation resilience.[12] Goals that are relevant to installations include reducing energy consumption or increasing energy efficiency, reducing pollution (e.g., greenhouse gas emissions), and achieving long-term energy security. Although some of these objectives, and the projects that support them, can and do increase installation resilience more broadly, they do not focus primarily on disruptions or costs from *extreme weather events* specifically.

Projects that support these other objectives (i.e., not targeted toward extreme weather events, per se) include installing microgrids, solar fields, energy storage batteries, and water treatment plants, as well as

---

[11] The depth of floodwaters and flow, in the case of non-coastal inundation, also drives damage cost.

[12] DoD, 2021; Office of the Assistant Secretary of the Army for Installations, Energy, and Environment, *United States Army Climate Strategy*, Department of the Army, February 2022; Office of the Assistant Secretary of the Navy for Energy, Installations, and Environment, *Department of the Navy Climate Action 2030*, Department of the Navy, May 2022; Department of the Air Force, *Department of the Air Force Climate Action Plan*, October 2022.

making buildings more energy efficient by attaining Leadership in Energy and Environmental Design (LEED) certification, adding water reclamation capability, and installing building control systems.[13]

Installations also have broader environmental compliance and stewardship goals. These include conserving natural resources, preserving or improving environmental quality, environmental restoration, and protecting local wildlife habitats.[14] The Services, and sometimes individual regions and installations, have policies and plans to achieve these goals as well.[15]

Finally, installations have goals that do not involve climate change or extreme weather events at all. These include installation safety and security and overall quality of life for Service members and their families.[16]

Each of these goals provides different lenses through which the installation and its facilities can be viewed. Using only one lens to view and prioritize investments or other actions would lead to an incomplete picture. Even limiting the perspective to environmental or climate-related goals and concerns would still result in competing objectives and difficult trade-offs.

Although the analytic and policy content in this report is focused purely on financial concerns (and only those related to extreme weather events), to clarify the specifically financial trade-offs, we do not advocate that DoD or the Services use only one lens to make these decisions. Any insights gleaned from the analysis in this report should be considered in the broader decision context.

In Chapter 5, we compile key findings from our analysis and close with a set of recommendations for next steps.

---

[13] See, for example, Office of the Assistant Secretary of the Army for Installations, Energy, and Environment, 2022, pp. 5–9; and Office of the Assistant Secretary of the Navy for Energy, Installations, and Environment, 2022, pp. 20–23.

[14] See, for example, DoD Environment, Safety & Occupational Health Network and Information Exchange, "2021 Secretary of Defense Environmental Awards," webpage, undated.

[15] See, for example, Army Regulation 200-1, *Environmental Quality: Environmental Protection and Enhancement*, Department of the Army, December 13, 2007. See also Commander Navy Region Northwest, "Environmental Stewardship and Compliance," webpage, undated; and Air Force Civil Engineering Center, "Environmental Management," webpage, undated.

[16] Quality of life for Service members and their families is a high priority for DoD and the Services. See, for example, DoD, "DoD Announces Immediate and Long-Term Actions to Help Strengthen the Economic Security and Stability of Service Members and Their Families," news release, September 22, 2022; U.S. Army, "Army Quality of Life," webpage, undated.

CHAPTER 5

Conclusions and Recommendations

In this chapter, we provide a high-level summary of the results described in greater depth in Chapter 3; share key findings related to the analytic approach, as well as data needs and availability; and conclude with a few recommended options for a way forward.

Overview of Illustrative Analytic Results

We have discussed elsewhere in the report the inherent limitations in this analysis. Thus, this overview should be considered applicable only to the specific installations we analyzed in Chapter 3.

Assuming that the costs of implementing a resilience option can be evenly spread over the option's lifetime (a bounding assumption that makes the option look much more attractive than it would under more-realistic financing schemes),

- For coastal flooding events, 6 feet of building-level floodproofing shows promise as a resilience option. The 9-foot floodwall option pays off only if coastal floods increase in frequency.
- Alternatives for storm surge events are less promising; only the 6-foot building floodproofing option yielded cost savings under current storm surge event probabilities for a subset of the coastal installations included in our analysis.
- For hurricane winds, we find that none of the considered resilience options provides positive annual savings until hurricane wind events become at least twice as frequent as they are currently. Across installations, window upgrades appear to pay off sooner than roof upgrades. Similarly, across installations, window upgrades are less costly to implement than roof upgrades, making the savings-yielding frequency multiplier smaller for window upgrades, even though roof upgrades offer more protection from extreme hurricane winds.

Assuming instead that the cost of implementing the 6-foot building floodproofing must be paid over a portion of the option's lifetime (over half of its lifetime or all within the first year) unsurprisingly results in the frequency multipliers increasing in most cases, i.e., events of interest would need to occur more frequently than they have historically for the resilience options to yield cost savings.

- For instance, if the full cost of the resilience option must be paid in the first year, coastal floods must become at least twice as frequent for 6 feet of building floodproofing to yield savings at any installation.
- However, we find that 6 feet of building floodproofing yields savings under current event probabilities for storm surge for at least one installation—NAVSUPPACT Hampton Roads—even if the implementation cost must be paid in the first year. This result is largely attributed to the relatively low cost of implementing this option at that installation, as well as the significant amount of avoided damage the resilience option offers the installation against storm surge events. The 6 feet of building floodproof-

ing would also protect NAVSUPPACT Hampton Roads against coastal flood damage, highlighting the importance of resilience options that protect against multiple hazards.

The results presented in this report are illustrative only and should not be directly used to inform resilience investment decisions. Additional analyses are needed before applying the insights of this study to DoD decisions regarding where and how much to spend on installation resilience to climate-driven hazards.

Key Findings

We have demonstrated that Hazus or a similar tool could be used to begin to understand the value of investments in installation resilience to climate-driven hazards. That said, available data to support analyses of the sort we present in this report are limited. For instance, although central DoD databases are a credible place to take stock of asset inventories at installations (e.g., numbers and types of assets), no single database offers comprehensive, high-quality data related to asset-level features, such as geolocations, PRV, other measures of an asset's financial value, or resilience options that are already in place. Conducting an analysis of the sort we present in this report requires drawing on a wide variety of data sources, crosswalking (and as discussed in the appendixes, re-coding) the data as needed to piece together all critical information to conduct the analysis.

Considering these data and modeling challenges, one alternative to modeling and simulation–based analyses is to learn from real-world events as they occur and from historical installation storm damage data. Extrapolating outcomes from these events to other contexts, whether in terms of damage incurred or the effectiveness of the resilience options used, could serve as a basis for comparing averted disaster damage costs with the cost of relevant resilience options. The Services are taking steps in this direction. U.S. Army Installation Management Command's storm damage dataset and Air Force Installation and Mission Support Center's storm damage tracker are good examples of data tracking efforts that could contribute to analysis that uses the historical record as a basis. **This approach has a downside as well: Data might not have been captured for some installations, data might be difficult to extract from installation data sources, and many DoD installations might not have been affected by enough extreme weather events to have sufficient data points to offer.** Where historical data are unavailable, assumptions would have to be made about the way in which a storm of a given type would affect facilities at location B, having occurred only at location A.

Recommended Options for a Way Forward

There are three broad ways in which the analysis presented in this report could be furthered—two apply to the **enterprise level** (DoD, OSD, or Service) and one applies in a more decentralized manner to the **installation level**.

At the enterprise level, the Hazus model, the cost factor data, and other non-DoD data and methods used in our approach need to be validated, verified, and customized for DoD purposes before being used for any decisionmaking purposes. Alternatively, investment could be made to develop tools tailored for DoD purposes.

Another enterprise-level step would require augmenting centrally available data with installation-provided data to conduct the analysis. Even if data gaps were fully addressed, the resource-intensiveness of a modeling and simulation–based exercise of the sort we undertook should be a key consideration in deciding whether to replicate such an exercise across the enterprise.

For decentralized, installation-level analysis, key enablers include guidance on evaluating resilience options; ensuring access to, and familiarity with, analytic tools, such as Hazus; and using mechanisms to track and record storm damage data. The same caveat regarding resource-intensiveness would apply here. Where installations have a key advantage is that they tend to have better knowledge of local assets, including information on the current level of resilience. Where installations would be similarly positioned to a Service or DoD would be in characterizing the effects of exposure to hazards of interest and in anticipating the effectiveness of resilience options that might be put into place.

Regardless of which of these approaches is pursued, OSD could play a role by

- issuing broad guidance to the Services about how to evaluate resilience options
- taking steps to improve the quality (e.g., consistency, compatibility) of RPAD and installation geospatial data
- establishing standard mechanisms through which installations can track, record, and report on data related to storm damage
- improving the familiarity of Service- or DoD-level personnel with analytical tools for estimating disaster-related damage.

Hazus Technical Documentation

This appendix provides more detail on how we ran Hazus. We refer the user to FEMA's online technical documentation and add only project-specific items here.[1]

Hazard Calculations (General Approach)

We used Hazus to calculate building damage for three types of hazards: hurricane winds, coastal floods, and storm surge. Hazus is a computer program developed by FEMA that allows users to simulate natural disasters and evaluate damage under various conditions.[2] The program, although not designed specifically for DoD use, allows great flexibility in scenario building to calculate the damage from events of different magnitudes and return periods.[3] Hazus could, for example, incorporate such information as hurricane probabilities and DEMs to define a hurricane's wind and coastal flooding damage.

For each hazard type, we followed a similar process in Hazus:

1. Hazus does not contain military installations in its default building set, first we uploaded the buildings (as user-defined facilities) into Hazus.
2. We opened a study region that is usable for either hurricanes or floods.
3. We ran the scenario (with specific steps varying as a function of the hazard type).

Step 1: Inputting Building Characteristics

Hazus does not contain military installations in its default building set.[4] Thus, first we uploaded the buildings (as user-defined facilities) into Hazus. This step involved (1) preparing the input data, and (2) uploading the data into Hazus.

- **Preparing the input data.** The first step involved creating a crosswalk between the RPAD data and the installation GIS data. The RPAD data contained information on building value, age, number of stories,

---

[1] FEMA, undated-b.

[2] FEMA, "Hazus," webpage, undated-a.

[3] Hazus was developed using historical damage data for basic community, business, and residential building types and sizes, and it does not include particular building types or larger-sized buildings that are important to military missions. For example, the U.S. Army has larger-sized maintenance buildings to repair bigger combat vehicles (because more bay space and larger cranes are needed to repair a tank compared with a wheeled tactical vehicle), and this type of building is not represented in Hazus.

[4] More information about Hazus's default inventory can be found at FEMA, *Hazus Inventory Technical Manual: Hazus 4.2 Service Pack 3*, February 2021b.

and square footage, but they did not include any information about the latitude and longitude of buildings within an installation. Geographic location is a required input in Hazus. To facilitate the crosswalk, we used each building's unique RPUID, which is present in both RPAD and GIS data. We used each building's centroid to construct our latitude and longitude variables.

Next, we categorized each building within each installation based on the building type classification required by Hazus. Each Hazus building type has different attributes, and the damage functions in Hazus depend on the building type. Thus, the losses associated with a natural disaster will vary based on the characteristics of the building. Table A.1. displays the PRV for each building type across the Bragg and Norfolk clusters. Across installations, the Hazus building type *concrete, engineered commercial building, high-rise (6+ stories)* (CECBH) tends to have the highest total PRV.[5] For some buildings, we were able to link RPAD data to one of the hurricane building types in Hazus, assuming the category *concrete, engineered commercial building, low-rise* (CECBL) as a default.[6]

- **Uploading the data into Hazus.** We then uploaded the buildings as user-defined facilities, following the Hazus comprehensive data management system technical manual, a program specifically designed for loading user-defined facilities.[7] Because we lack data on DoD building attributes, including existing hazard mitigation investment, we allowed Hazus to populate information with the default building first-floor height of 1 foot.

TABLE A.1

Distribution of Hazus Building Types Across Installations

Cluster	Installation	CECBH	CECBL	CECBM	CERBL	CERBM	CERBH	SPMBM	SPMBS	Total
Bragg	Fort Bragg	7.3	9,668.6	2,467.5	4,663.3	1,687.0	0.0	431.9	36.6	18,962.1
Norfolk	JBLE	9.2	3,388.7	1,036.2	191.9	659.1	60.0	0.0	0.0	5,345.2
	NWS Yorktown	92.1	1,207.9	7.7	190.9	97.7	0.0	0.0	0.0	1,596.4
	NAVSTA Norfolk	1,099.9	3,476.1	885.7	798.4	336.6	300.0	0.0	0.0	6,896.7
	NAVSUPPACT Hampton Roads	341.9	119.5	921.6	2.2	68.8	0.0	0.0	0.0	1,454.1
	NAVSUPPACT Norfolk NSY	46.5	1,651.3	625.7	170.8	138.4	91.3	0.0	0.0	2,724.0

SOURCE: Authors' calculations using RPAD and Hazus data.

NOTE: Additional Hazus building types for the flood and hurricane analysis are as follows: CECBM = concrete, engineered commercial building, mid-rise (3–5 stories); CERBL = concrete, engineered residential building, low-rise (1–2 stories); CERBM = concrete, engineered residential building, mid-rise (3–5 stories); CERBH = concrete, engineered residential building, high-rise (6+ stories); SPMBM = steel, pre-engineered metal building, medium; SPMBS = steel, pre-engineered metal building, small.

---

[5] Hazus also requires a field that describes each building's occupancy. However, Hazus uses this field only to estimate damage costs of a building's contents, which we do not assess in this report.

[6] Hurricane building types are defined in FEMA, 2021b, Section 4.2.3.

[7] FEMA, *Hazus Comprehensive Data Management System (CDMS) User Guidance*, April 2019.

Limitations

Note that this study is limited in several ways:

- This study included only building data; because Hazus cannot take other types of inputs (e.g., roads, communication lines), such inputs would need to be costed out separately. The implication of this limitation is that any area of effect mitigation (such as a floodwall) would be even more attractive (from a cost-benefit analysis) than our study shows.
- All buildings had to be converted to a building within Hazus's building types; some buildings (such as radars) could not be mapped, and others might have been mapped incorrectly. This means that the analysis conducted should be considered as an uncertain value.
- We used only building cost (PRV) and set building contents to zero. The implication of this is that any mitigation would be even more attractive (from a cost-benefit analysis) than our study shows.

Step 2: Defining Study Regions

Hazus is computationally expensive, so we chose the minimum study region size.

- First, we considered geographic scope minimization. To do this, we viewed the input facility data for each installation and determined the minimum number of counties that would encompass the data.
- Second, we considered resolution minimization. The coarsest resolution at which we were able to conduct analysis without loss of resolution was the census tract level (flooding and wind both vary as a function of this). We considered a finer resolution but determined our study region's topography was homogeneous enough that this finer resolution did not affect results.
- Third, given the way the flood maps were calculated, we were able to combine all the Norfolk installations (NWS Yorktown; NAVSTA Norfolk; NAVSUPPACT Hampton Roads; NAVSUPPACT Norfolk NSY; and JBLE).

Given this, we used the two defined Hazus study regions (see Table A.2). We then followed the standard

TABLE A.2
Hazus Study Regions

Study Region	Counties	Hazards Examined
Fort Bragg	Cumberland County, N.C.	Hurricane winds
Norfolk installations	Chesapeake County, Va.; Hampton County, Va.; James City County, Va.; Newport News, Va.; Norfolk County, Va.; Poquoson, Va.; Portsmouth, Va.; York, Va.	Hurricane winds, storm surge, coastal flooding

NOTE: This table shows the counties used to define the Hazus study regions and the hazards examined within each study region.

Hazus online documentation to create these study regions.[8]

---

[8] FEMA, 2022a; FEMA, April 2022b.

Step 3a: Hurricane Winds

Defining the Scenarios

Hazus has a default setting for calculating probabilistic hurricane wind speeds as a function of return period (see Section 4 of the *Hazus 5.1 Hurricane Model User Guidance*).[9] We used these default settings, resulting in the construction of wind maps as a function of return period in Figures A.1–A.14.

FIGURE A.1

Hurricane Winds, Ten-Year Return Period, Fort Bragg

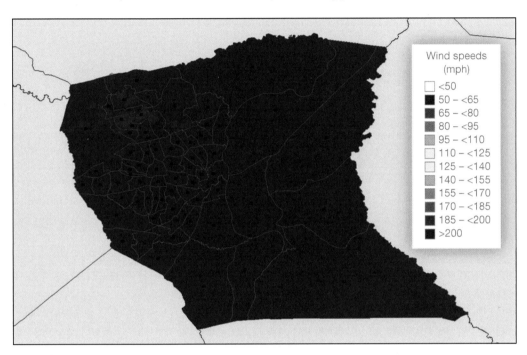

SOURCE: Authors' calculations using RPAD and Hazus data.

---

[9] FEMA, 2022b, Section 4.

FIGURE A.2

Hurricane Winds, 20-Year Return Period, Fort Bragg

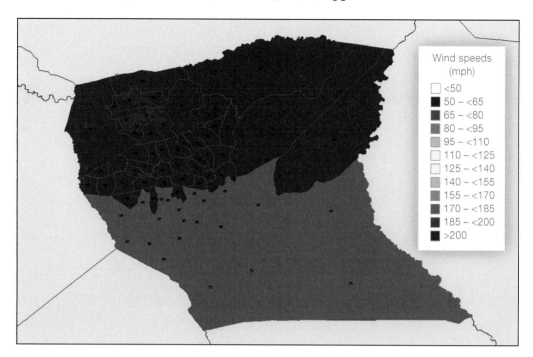

SOURCE: Authors' calculations using RPAD and Hazus data.

FIGURE A.3

Hurricane Winds, 50-Year Return Period, Fort Bragg

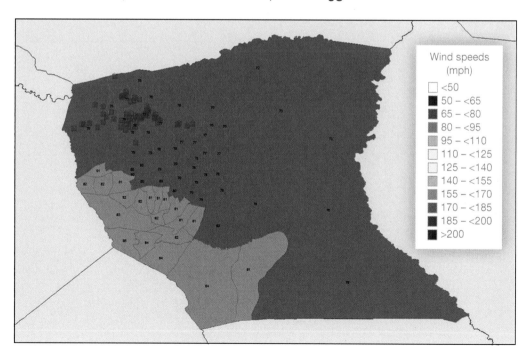

SOURCE: Authors' calculations using RPAD and Hazus data.

FIGURE A.4

Hurricane Winds, 100-Year Return Period, Fort Bragg

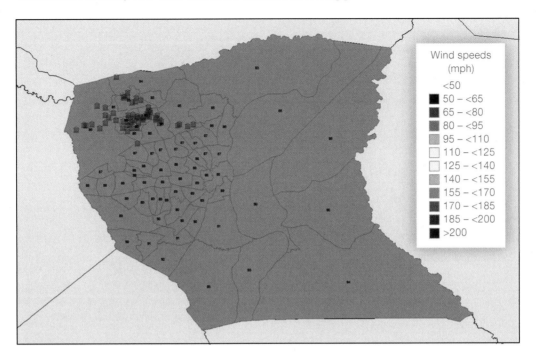

SOURCE: Authors' calculations using RPAD and Hazus data.

FIGURE A.5

Hurricane Winds, 200-Year Return Period, Fort Bragg

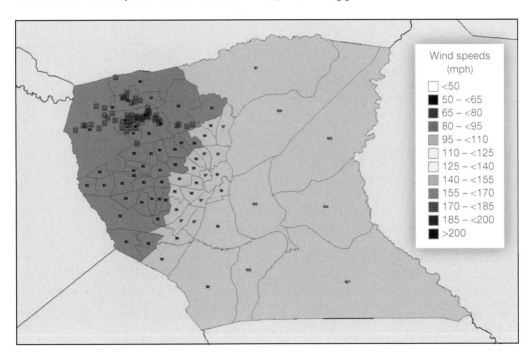

SOURCE: Authors' calculations using RPAD and Hazus data.

FIGURE A.6

Hurricane Winds, 500-Year Return Period, Fort Bragg

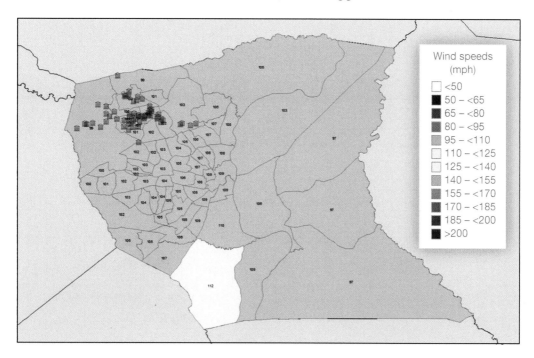

SOURCE: Authors' calculations using RPAD and Hazus data.

FIGURE A.7

Hurricane Winds, 1,000-Year Return Period, Fort Bragg

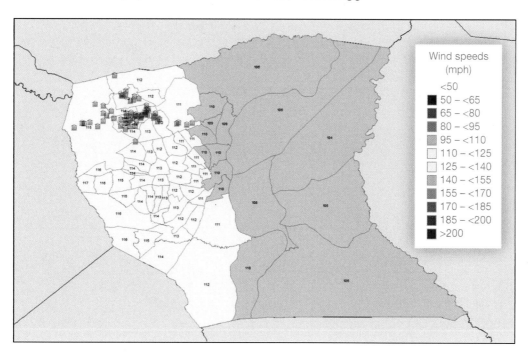

SOURCE: Authors' calculations using RPAD and Hazus data.

FIGURE A.8

Hurricane Winds, Ten-Year Return Period, Norfolk Installations

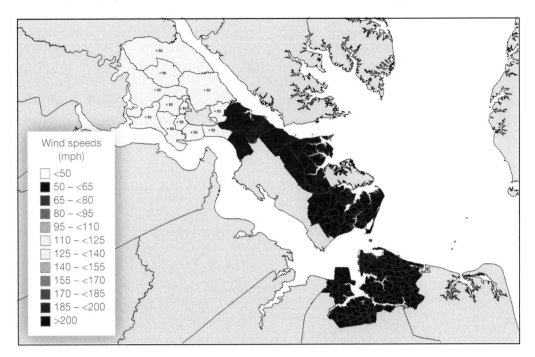

SOURCE: Authors' calculations using RPAD and Hazus data.

FIGURE A.9

Hurricane Winds, 20-Year Return Period, Norfolk Installations

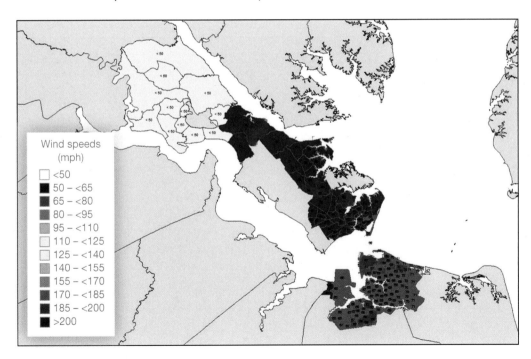

SOURCE: Authors' calculations using RPAD and Hazus data.

FIGURE A.10

Hurricane Winds, 50-Year Return Period, Norfolk Installations

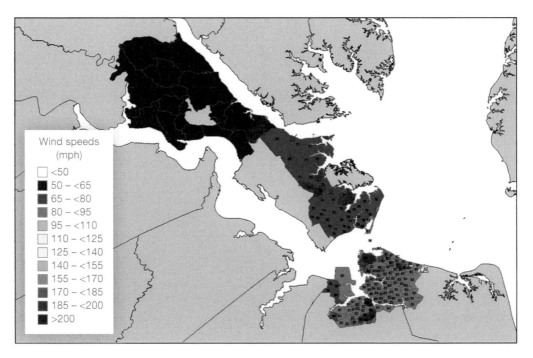

SOURCE: Authors' calculations using RPAD and Hazus data.

FIGURE A.11

Hurricane Winds, 100-Year Return Period, Norfolk Installations

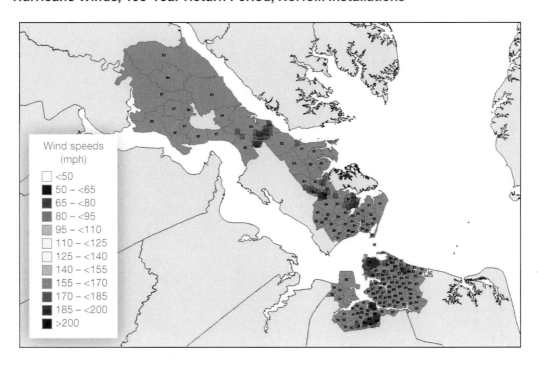

SOURCE: Authors' calculations using RPAD and Hazus data.

FIGURE A.12

Hurricane Winds, 200-Year Return Period, Norfolk Installations

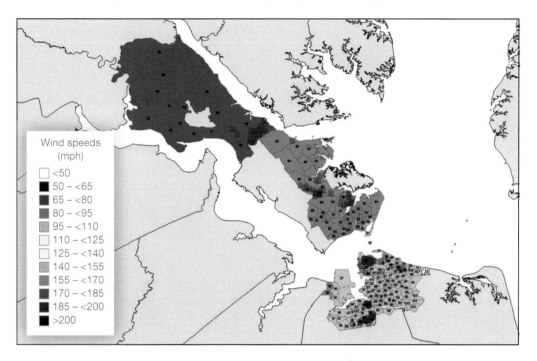

SOURCE: Authors' calculations using RPAD and Hazus data.

FIGURE A.13

Hurricane Winds, 500-Year Return Period, Norfolk Installations

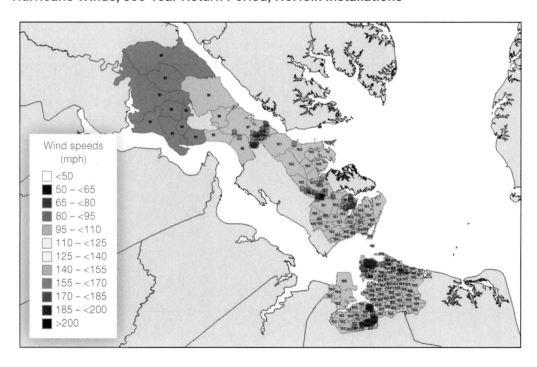

SOURCE: Authors' calculations using RPAD and Hazus data.

FIGURE A.14

Hurricane Winds, 1,000-Year Return Period, Norfolk Installations

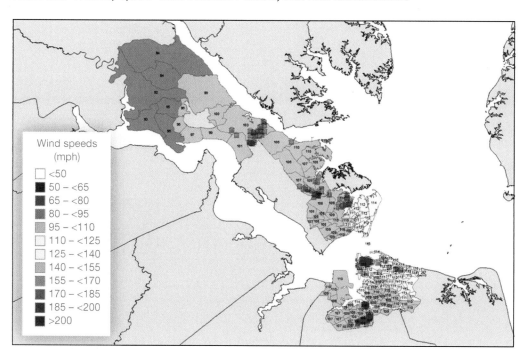

SOURCE: Authors' calculations using RPAD and Hazus data.

Running Analysis

We ran two baseline simulations: Simulation A made no change to Hazus's default building characteristics (which are an average from census tracts with similar climate zones), and simulation B had all resilience options turned off. We used simulation A as the baseline case in the main report. We then ran three retrofit scenarios. Table A.3. shows the input characteristics used for these scenarios. Building characteristics were updated in Hazus using the menus *Inventory -> General Building Stock* (to change the characteristics) and *Inventory -> User Defined Facilities* (to apply the changes to the buildings). Note that information was not available for each column for all hurricane building types, as defined in Table A.3.

After updating the building characteristics, we followed the steps in the FEMA hurricane user manual to run the scenarios.

Limitations

A limitation of this approach is that we did not combine wind and storm surge into one model. Although Hazus has this potential, we needed to separate the wind and storm surge building datasets in such a way that did not allow us to recombine them.[10] Because wind and storm surge mitigations were considered in isolation, we considered this to be sufficient for this study. A future study could combine these datasets, and then run Hazus with the combination.

This analysis considers hurricane wind speeds only. A more complete analysis would consider winds from all sources (likely making certain wind speed events more frequent). This would mean that if any mitigation

---

[10] FEMA, 2022a, Section 11.

TABLE A.3

Hurricane Wind Mitigation Scenarios Run in Hazus

Scenario	Roof Cover Type	Window Area	Shutters	Wind Debris	Metal Roof Deck Attachment	Terrain
Base case (Hazus default)	Single-ply membrane	No change to assumptions for area	No	No change to assumptions for area	Standard	No change to assumptions for area
Retrofit roofing system only	Built-up roof[a]		No	No change to assumptions for area	Superior[a]	No change to assumptions for area
Retrofit exterior windows only	Single-ply membrane	No change to assumptions for area	Yes[a]	No change to assumptions for area	Standard	No change to assumptions for area
Retrofit roofing system and exterior windows	Built-up roof[a]	No change to assumptions for area	Yes[a]	No change to assumptions for area	Superior[a]	No change to assumptions for area

SOURCE: Authors' calculations using Hazus data.

NOTE: For all inputs, we used Hazus's assumptions based on input census data.

[a] Green shading indicates an enacted mitigation that was an improvement over baseline.

solutions were found to be attractive (in a cost-benefit analysis sense) for hurricane wind speeds, those solutions would likely be even more attractive when considering winds from all sources.

Step 3b: Coastal Flooding

Defining the Scenarios

Hazus can calculate coastal flooding as a function of return period (see Section 6 of the *Hazus 5.1 Flood Model User Guidance*).[11] We used Hazus to identify the DEMs and used Hazus's default coastal scenario method (Section 6.4.1.2. of the *Hazus 5.1 Flood Model User Guidance*).[12] This method takes as an input the 100-year event still water levels (no waves). We used SERDP's and ESTCP's Regional Sea Level Rise Change Scenarios database to define these levels, resulting in 8.5-foot, 100-year still water levels.[13] Otherwise, we accepted default settings, and then delineated the floodplain following guidance in the Hazus manual.

A series of sample flood maps (for the 500-year return period) is shown in Figures A.15–A.17. The first map shows the DEM and buildings (in black) in the study region. The second set of maps shows two views of the 500-year coastal flood (as calculated by Hazus) and how it overlays on the buildings in the study region. Other return period maps show a similar pattern.

Running the Analysis

We ran one baseline simulation with no change to Hazus's default building characteristics (the first floor is elevated to 1 foot, and there is no flood protection). We then calculated damage with mitigation.

---

[11] FEMA, 2022a, Section 6.

[12] FEMA, 2022a, Section 6.4.1.2.

[13] Department of Defense Strategic Environmental Research and Development Program and Environmental Security Technology Certification Program, undated.

FIGURE A.15
Digital Elevation Map, Norfolk Installations

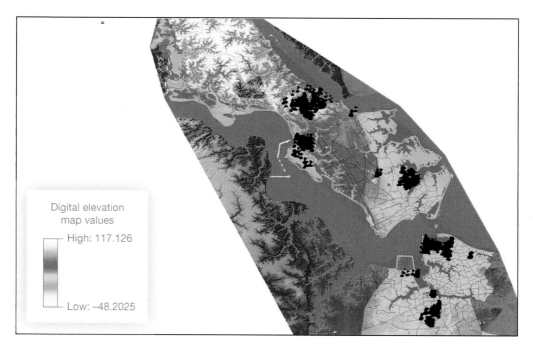

SOURCE: Authors' calculations using RPAD and Hazus data.

FIGURE A.16
500-Year Return Period Flood Map, Without Climate Change

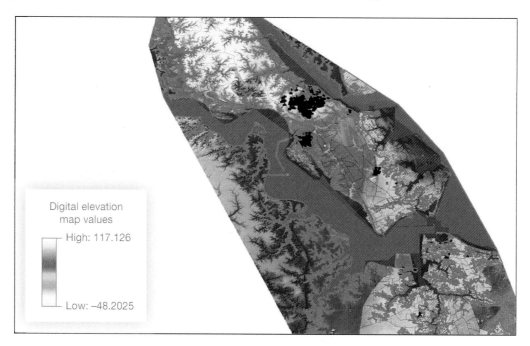

SOURCE: Authors' calculations using RPAD and Hazus data.
NOTE: Only installations above the flood waters are shown; compare with Figure A.15 to identify flooded locations.

FIGURE A.17

500-Year Return Period Flood Map, with Climate Change

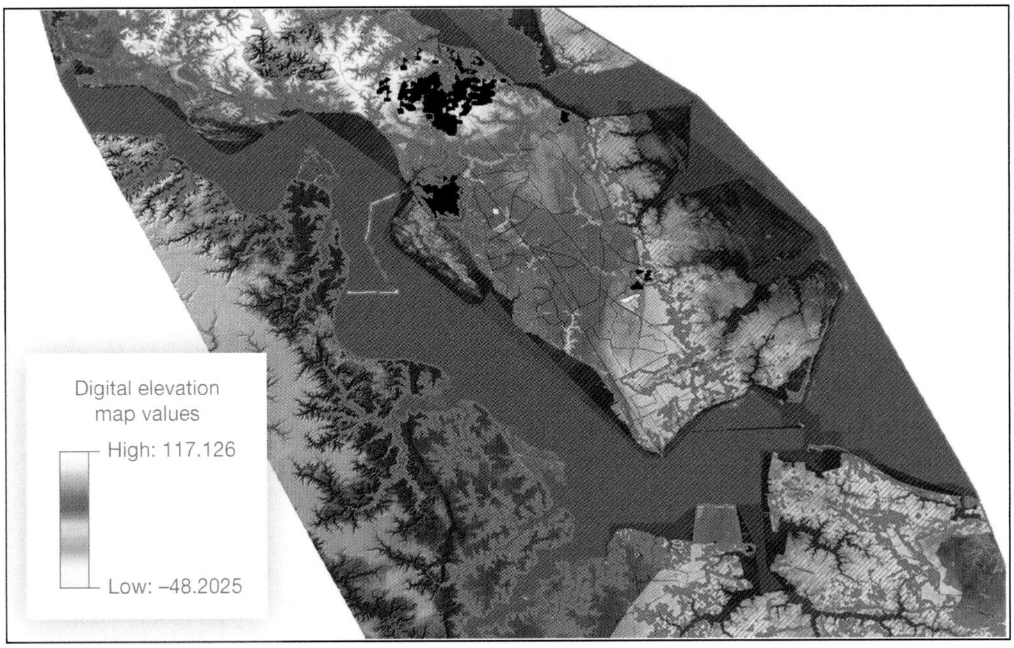

Digital elevation
map values

High: 117.126

Low: −48.2025

SOURCE: Authors' calculations using RPAD and Hazus data.
NOTE: Only installations above the flood waters are shown; compare with Figure A.15 to identify flooded locations.

Post-Processing

We post-processed the elevation data into four scenarios for each elevation level, as shown in Table A.4.

Limitations

Regarding hazard probability, we have two main limitations. First, we are limited to the coastal flooding models within Hazus, which are likely not as complete as a full hydraulic model.

TABLE A.4

Coastal Flood Scenarios

Scenario	Scenario Description	Calculation Method
A	No mitigation (by return period)	Use output in Hazus as is
B	6-foot floodproofing (by return period)	Elevate buildings by 6 feet, then use output in Hazus as is
C	9-foot floodwall (by return period)	Look up scenario level tides. Assume zero damage when floodwall is not overtopped, and assume floodwall overtops (use damage from A) when flooding value is higher than 9 feet for that return period
D	6-foot floodproofing and 9-foot floodwall (by return period)	Look up scenario level tides. Assume zero damage when floodwall is not overtopped, and assume floodwall overtops (use damage from B) when flooding value is higher than 9 feet for that return period

For the mitigations, we are limited by the available building data. Although we were able to obtain location data for individual facilities, we were unable to obtain data on their first-floor heights. Thus, it is unclear how many buildings might already be mitigated against flood.

Step 3c: Storm Surge

Defining the Scenarios

Because Hazus has pre-defined probabilistic hurricane wind speeds as a function of return period (see "Step 3a: Hurricane Winds" in Appendix A), we chose to match storm surge to these values.

First, recall that Hazus sampled from all types of possible hurricanes to create wind maps as a function return period, as opposed to creating a wind map from a single hurricane. Thus, we needed to similarly sample all types of possible storm surges to create the storm surge maps. The National Hurricane Center has created these maps, called Maximums of Maximum Envelope of Open Water (MOMs), as a function of the Saffir-Simpson scale.[14] Therefore, we downloaded the Norfolk MOMs for above ground level (AGL).[15] We then imported these maps into Hazus as an Environmental Systems Research Institute grid, following Section 12.1 of the FEMA flood model guidance.[16] Figures A.18–A.21 show the resulting grids.

---

[14] National Hurricane Center and Central Pacific Hurricane Center, undated-b.

[15] National Hurricane Center and Central Pacific Hurricane Center, undated-b. The data are also available as above datum, but this requires more computations within Hazus, and thus we used AGL.

[16] FEMA, 2022a, Section 12.1.

FIGURE A.18

AGL MOM for Hurricane, Saffir-Simpson Category 1

SOURCE: Authors' calculations using National Hurricane Center and Hazus data.
NOTE: AGL = above ground level.

FIGURE A.19

AGL MOM for Hurricane, Saffir-Simpson Category 2

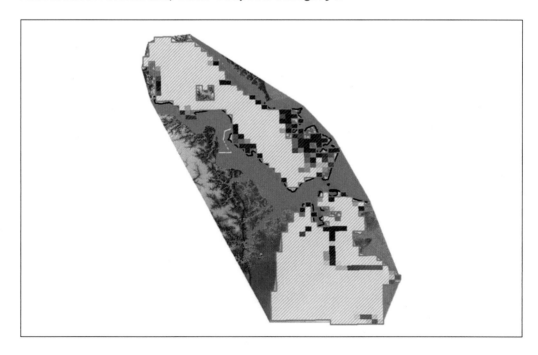

SOURCE: Authors' calculations using National Hurricane Center and Hazus data.

FIGURE A.20
AGL MOM for Hurricane, Saffir-Simpson Category 3

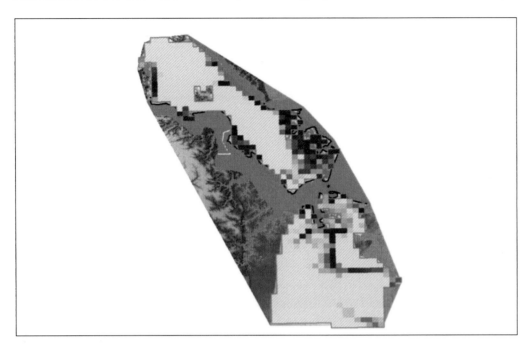

SOURCE: Authors' calculations using National Hurricane Center and Hazus data.

FIGURE A.21
AGL MOM for Hurricane, Saffir-Simpson Category 4

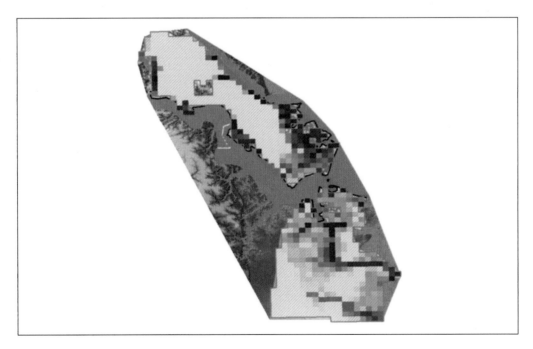

SOURCE: Authors' calculations using National Hurricane Center and Hazus data.

To map these scenarios to a return period, we used the Hazus wind maps (Figures A.1–A.14) to identify the Saffir-Simpson scale for each installation as a function of return period. We then mapped the scale to the corresponding storm surge MOMs.[17] This conversion is displayed in Tables A.5–A.7.

Running the Analysis

We followed the same steps as in the coastal flood analysis.

Post-Processing

We followed the same steps as in the coastal flood analysis.

Limitations

The limitations of our storm surge analysis are the same as the limitations for our hurricane and coastal flooding analyses. Additionally, we did not combine wind and storm surge into one model. Although Hazus has this potential, we needed to separate the wind and storm surge building datasets in such a way that did not allow us

TABLE A.5
Definition of Saffir-Simpson Wind Speeds

Storm Surge Run	Hurricane Category	Wind Speeds (mph)
AGL 1	1	74–95
AGL 2	2	96–110
AGL 3	3	111–129
AGL 4	4	130–156

SOURCE: Features information from National Hurricane Center and Central Pacific Hurricane Center, undated-a.

TABLE A.6
Maximum Winds in Hazus at Each Installation

Site	Return Period						
	10 Years	20 Years	50 Years	100 Years	200 Years	500 Years	1,000 Years
NWS Yorktown	50	52	65	85	79	98	103
JBLE	52	62	74	84	90	103	110
NAVSTA Norfolk	54	67	81	86	96	104	112
NAVSUPPACT Hampton Roads	55	70	85	89	99	106	113
NAVSUPPACT Norfolk	58	72	83	92	101	108	115

SOURCE: Authors' calculations using Hazus data.

NOTE: Table values represent wind speeds in mph.

---

[17] National Hurricane Center and Central Pacific Hurricane Center, "Saffir-Simpson Hurricane Wind Scale," webpage, undated-a.

TABLE A.7

Estimated Return Period (in Years) for Each Storm Surge Run at Each Installation

	NWS Yorktown	JBLE	NAVSTA Norfolk	NAVSUPPACT Hampton Roads	NAVSUPPACT Norfolk
AGL 1	75	50	35	30	25
AGL 2	470	340	200	170	145
AGL 3	> 1,000	> 1,000	940	860	710
AGL 4	> 1,000	> 1,000	> 1,000	> 1,000	> 1,000

SOURCE: Authors' calculations using Hazus data.

NOTE: This table displays estimated return periods as a function of Tables A.5 and A.6.

to recombine them.[18] Because wind and storm surge mitigations were studied in isolation, we considered this to be sufficient for this study. A future study could combine these datasets and then run Hazus.

---

[18] FEMA, 2022a.

Wildfire Technical Documentation

Wildfire Calculations (General Approach)

As mentioned, Hazus does not include a wildfire model. In addition, the types of mitigations typically used to reduce wildfire damage are area-of-effect type efforts. Thus, we tried a different approach for wildfires and ultimately decided not to include the results from this analysis.

First, we assumed that if a building is exposed to wildfire, it will be 100 percent damaged. Then, for the Colorado installations, we did the following:

1. Identified the probability of exposure to wildfires for each building.
2. Defined the resilience options (reflecting state-wide guidance).
3. Because step 1 resulted in all area-of-effect mitigations that prevented 100 percent of damage, we simplified the risk equation. We thus used steps 1 and 2 to define the risk.

Step 1: Identify the Probability of Exposure to Wildfires for Each Building

According to informal discussions with subject-matter experts, there was a possibility that the Colorado installations might already have superior wildfire protections (such as vegetation management) in place, and thus might not be exposed to wildfires in the future. Given this possibility, we first checked whether, in the recent past (1984–2020), the Colorado installations had been affected by wildfire. Figures B.1 and B.2 show historical reanalyzed data on wildfires in Colorado from the Monitoring Trends in Burn Severity (MTBS) dataset.[1] Figures B.1 and B.2 show the installation locations and the MTBS burned area extents from 1984–2020. Given Colorado State Forest Service guidance to employ vegetation management of 100 feet (which is an upper limit for the distance embers might be able to float away from the fire), we draw a 100-foot buffer around the burned areas to indicate the zone at risk.[2] We see that multiple installations are collocated with burned areas, suggesting that there is a need to continue considering wildfire threats.

Unfortunately, current wildfire modeling tools are very limited and are unsuitable for use in the DoD context. Considering this limitation, we used the U.S. Department of Agriculture's national burn probability data, available from USFS, which defines burn probability as the "annual probability of wildfire burning in a specific location."[3]

---

[1] MTBS, "Direct Download,"datasets, undated. Note that MTBS indicates several types of wildfires, including normal wildfires or managed wildfires (e.g., proscribed burns, agricultural burns).

[2] Colorado State Forest Service, "Protect Your Home & Property from Wildfire," webpage, undated.

[3] Short et al., 2016.

FIGURE B.1

**MTBS Burned Area Extents, Without Counties
(1984–2020)**

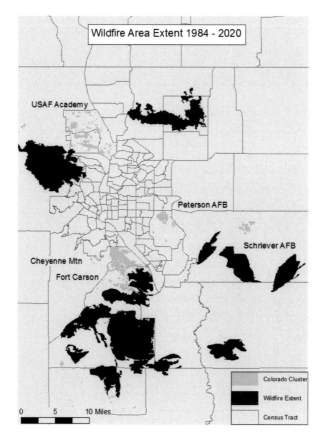

SOURCE: RAND geospatial analysis using RPAD and MBTS data.

FIGURE B.2

MTBS Burned Area Extents, with Counties (1984—2020)

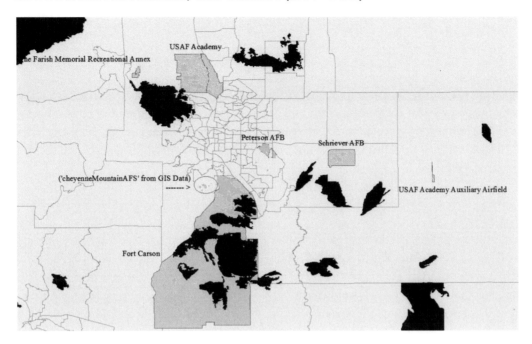

SOURCE: RAND geospatial analysis using RPAD and MBTS data.

Step 2: Define Resilience Options

The primary method typically used to protect against wildfires is the creation of defensible space.[4] The general premise is that assets will be protected by the inability of the embers to reach them during a fire. Given Colorado State Forest Service guidance to employ vegetation management of 100 feet (an upper limit for the distance the embers might be able to float away from a fire), we use this distance in our resilience option.[5]

Other options are less preferred and are typically reserved for spaces where vegetation cannot be cleared.[6] For example, avoiding wood buildings; to our knowledge, the installations already do this, so that option would not make sense here. Another option is emergency operations, such as dumping water from helicopters, or fire retardants (foam or similar): These are emergency response measures, which are not considered in this study.

Step 3: Running the Analysis

As described in Chapter 2, for each resilience option, installation, and disaster type, we calculate the annualized cost of averted damage as

$$AvertedDamage = \sum_{\{i=1\}}^{m} \frac{\sum_{\{n=1\}}^{L}\left[\binom{L}{n}*EAP_i^n*(1-EAP_i)^{L-n}\right]*\left[BaseDamage - ResDamage\right]}{L} \quad (B.1)$$

---

[4] F. C. Dennis, "Fire-Resistant Landscaping," Colorado State University Extension, Fact Sheet No. 6.303, undated.

[5] Colorado State Forest Service, undated.

[6] Insurance Institute for Business & Home Safety, "Wildfire," webpage, undated.

We made two simplifying assumptions. First, because of the USFS dataset, we consider only one magnitude of wildfire. Thus, $m = 1$ and the EAP is what is displayed in the USFS choropleth map. As noted in the main text, the annual burn probability is converted into a return period as

$$\frac{1}{\max(\textit{burn probability})}$$

where the denominator is the maximum burn probability across all buildings within the installation.

Second, we assume that 100 percent of the building is damaged if an installation geographically intersects with a wildfire and that 0 percent of the building is damaged otherwise. As a result, our analysis focuses only on buildings located on so-called burnable land, and *BaseDamage* and *ResDamage* are either 0 percent or 100 percent of the PRV.

Limitations

This analysis is greatly limited by the lack of sophistication in existing wildfire models. The models themselves are highly uncertain, as are the resulting assumptions of flame height and how those heights harm buildings. It is likely that the models would be improved with better flame height representation, which would then allow incremental percentage damage (other than 0 or 100 percent). For example, the U.S. Department of Agriculture has a model that includes both national burn probabilities and treatment of flame heights, which might conceivably be linked to damage probabilities.[7] Similarly, it might be possible to consider interpretation of the *Wildfire Hazard Potential for the United States*, which links both wildfire probabilities and flame heights.[8] But improved engineering damage data (such as the damage curves within Hazus for wind and water) would be needed to build confidence in the strength of such relationships.

In addition, this analysis is limited by assuming just one resilience option—vegetation management. Other options include strong building codes, dumping water from helicopters, or use of fire retardants (foam or similar).

---

[7] See, for example, Short et al., 2016.

[8] Gregory K. Dillon and Julie W. Gilbertson-Day, "Wildfire Hazard Potential for the United States (270-m)," 3rd ed., Forest Service Research Data Archive, 2020.

Cost Model

This appendix includes additional detail on the methodology and data sources used to develop the cost model for the resilience strategies. The first three sections contain the unit costs used in the cost model, the sources used to obtain the unit cost data, and the details on what is included in the resilience cost estimate for each hazard considered in the analysis (hurricane winds, coastal flooding and storm surge, and wildfire). The fourth section includes descriptions of the cost factors incorporated into the cost model and the methodology for calculating annualized costs from total costs for each resilience strategy.

Hurricane Winds

The cost model to estimate resilience options for hurricane winds was based on retrofitting the roof and installing hurricane shutters. In our review of hazard resilience literature, we determined roof assemblies and exterior openings (e.g., windows) as vulnerable components of a building during hurricane winds.[1] Other components, such as the building structure or special structures (e.g., canopies or telecommunication towers), are also vulnerable to hurricane winds. However, estimating costs to implement resilience options for these items is difficult without a detailed cost model that uses higher-fidelity data.

Roof Replacement

The cost data used to estimate the roof assembly replacement was derived from RSMeans, a construction cost-estimating software. The roof assembly retrofit included replacement of the roof cover, insulation, deck, edges, flashing, and gravel stop.[2] We first collected unit costs for the roof assembly retrofit for different building types using the SFE tool in RSMeans. We then mapped the building asset types in RPAD using the facility analysis category to the building types in RSMeans to determine which unit cost to use in the roof retrofit cost estimates for a given building. For example, a *communications building* in the RPAD data is mapped to the *office, 2–4 story* building category in RSMeans. Additionally, the unit cost data from RSMeans are adjusted to account for regional variations using the CCI. The CCI is a feature in the RSMeans database to adjust construction cost data for a specific location in the United States. For a given installation, we used the closest metropolitan city near the installation. For example, we used the Durham, North Carolina, CCI for Fort Bragg. Table C.1 lists the unit costs by building type for the roof retrofit items for Fort Bragg.

The next step in developing the cost model was determining the amount of roof area and roof perimeter for each building. The asset-level data from RPAD included total building area and number of stories; how-

---

[1] Multi-Hazard Mitigation Council, *Natural Hazard Mitigation Saves: 2019 Report*, National Institute of Building Sciences, December 2019.

[2] Without knowing the condition of the roof, it was difficult to determine whether the roof deck would require a retrofit to withstand hurricane wind conditions. As a conservative approach, we included replacement of the roof deck in the roof assembly retrofit cost.

TABLE C.1

Unit Costs for Roof Retrofit by RSMeans Building Category for Fort Bragg

RSMeans Building Category	Roofing Cover, Insulation, and Decking ($/SF of Roof Area)	Roof Edges, Flashing, and Gravel Stop ($/LF of Roof Perimeter)
Apartment, 1–3 story	55.69	4.23
Bank	77.74	8.60
Bowling alley	65.52	10.01
Church	9.83	12.69
College, laboratory	65.52	12.69
Community center	65.52	10.01
Day care center	65.52	10.03
Factory, 1-story	65.52	10.01
Fire station, 1-story	65.52	10.01
Garage, repair	80.55	10.01
Garage, service station	9.83	10.74
Gymnasium	55.69	10.83
Hangar, aircraft	65.52	10.01
Hospital, 2–3 story	65.52	4.23
Jail	69.32	8.60
Laundromat	65.52	10.01
Library	55.69	4.69
Medical office, 1-story	65.52	12.69
Movie theater	69.57	8.62
Office, 1-story	65.52	9.74
Office, 2–4 story	65.52	4.23
Police station	65.52	5.00
Post office	65.52	10.01
Restaurant, fast food	65.52	8.81
School, vocational	65.52	6.00
Store, department, 3-story	65.52	3.34
Store, retail	65.52	12.69
Veterinary hospital	9.83	13.53
Warehouse	65.52	12.69
Warehouse, self-storage	78.43	8.76

ever, the number of stories was not consistently available for all building assets in RPAD. To calculate the roof area and the roof perimeter, we used the following equations:[3]

$$Roof\ Area\ = \frac{Building\ Area}{Number\ of\ Stories}$$

$$Roof\ Perimeter\ = \sqrt{\frac{Building\ Area}{Number\ of\ Stories}}.$$

For buildings where the number of stories was missing in RPAD, we estimated the number by extrapolating from the SFE for a given building type. Table C.2 includes estimated quantities of building features by RSMeans building category type.

Hurricane Shutter Installation

The data used to estimate costs for the installation of hurricane shutters on exterior windows were derived from RSMeans. Unit costs for hurricane shutters were based on cost per square foot of the exterior windows. Like the roofing retrofit costs, the unit costs were adjusted using the CCI to account for local variations in construction pricing.

The next step was to estimate the number of exterior windows for each building. The RPAD data did not include the number or area of exterior windows. To estimate the area of exterior windows, we estimated quantities of building features by RSMeans building category type. For example, a *communications building* in RPAD is categorized as an *office, 2–4 story*, which is estimated to have 20 percent of the exterior wall as windows. We calculated the exterior windows area using the following equation:

$$Exterior\ Window\ Area\ =\ (\%\ of\ Exterior\ Wall\ as\ Window)^*(\#\ of\ Stories)^*(Story\ Height)$$
$$^*(Building\ Perimeter).$$

Coastal Flooding and Storm Surge

Resilience options for coastal flooding and storm surge included both property-level and community-level approaches. The property-level approach was based on dry floodproofing of buildings, whereas the community-level approach was based on introducing a floodwall at the installation site. Other property-level and community-level resilience strategies to protect against coastal flooding exist. Examples include building elevation, improved drainage systems, and relocating assets out of flood prone areas. Our decision to include dry floodproofing and flood wall installation in the cost model was based on cost data availability and options in Hazus.

Dry Floodproofing

The unit cost data for dry floodproofing are based on FEMA literature.[4] We adjusted the unit cost data to account for inflation from 2002 to 2022, and to account for local variations in construction costs using the RSMeans CCI. The items included in dry floodproofing, and the respective unit costs for those items, are listed in Table C.3. We assume dry floodproofing provides protection up to 6 feet in height. Typically, walls

---

[3] The equation for the roof perimeter assumes a square building.

[4] FEMA, *Selecting Appropriate Mitigation Measures for Floodprone Structures*, FEMA 551, March 2007.

TABLE C.2

Estimates of Building Feature Quantities by RSMeans Building Category

RSMeans Building Category	Number of Stories	Story Height (ft)	Percentage of Exterior Wall as Windows (%)
Apartment, 1–3 story	3	10	20
Bank	1	14	20
Bowling alley	1	14	10
Bus terminal	1	14	0
Church	1	24	10
College, laboratory	1	12	9
Community center	1	12	20
Day care center	1	12	10
Factory, 1-story	1	20	25
Fire station, 1-story	1	14	22
Garage, repair	1	14	5
Garage, service station	1	12	20
Gymnasium	1	25	12
Hangar, aircraft	1	24	20
Hospital, 2–3 story	3	12	45
Jail	3	12	15
Laundromat	1	12	10
Library	2	14	10
Medical office, 1-story	1	10	29
Movie theater	1	20	20
Office, 1-story	1	12	31
Office, 2–4 story	3	12	20
Police station	2	12	22
Post office	1	14	20
Restaurant, fast food	1	10	25
School, vocational	2	16	56
Store, department, 3-story	3	16	56
Store, retail	1	14	14
Veterinary hospital	1	12	7
Warehouse	1	24	2
Warehouse, self-storage	1	12	6

TABLE C.3

Dry Floodproofing Unit Costs Based on National Average

Dry Floodproofing Item	Unit Cost ($)
Waterproofing concrete wall	5.64/SF
Acrylic latex wall coating	3.00/SF
Caulking or sealant	2.50/LF
Bentonite grout (below grade waterproofing)	20/LF

without reinforcement can withstand hydrostatic pressure from flood levels of up to 3 feet.[5] We assume that the wall construction for the buildings on an installation are reinforced and can withstand hydrostatic pressure to a flood level of up to 6 feet. The quantity in SF of waterproofing and acrylic latex wall coating used in the cost estimate is based on the equation below. The quantity of LF for caulking and bentonite grout is based on the building perimeter, which is calculated as follows using the same equation for the roof perimeter:

$$SF \ of \ Waterproofing \ and \ Acrylic \ Latex \ = \ (6 \ feet \ of \ protection)*(Building \ Perimeter).$$

Floodwall

The data sources used to develop the cost model for the floodwall are based on FEMA literature and a 2018 report by Jeroen Aerts.[6] The FEMA literature includes floodwall costs at 2-foot, 4-foot, and 6-foot levels of protection, whereas the 2018 report by Aerts includes floodwall costs for 13-foot and 24-foot levels of protection. These unit costs were adjusted for inflation and used as data points in a regression model to determine the unit cost of a floodwall for 9 feet of protection, as noted in Table C.4.[7]

To estimate the required length of the floodwall, we assumed that the floodwall would span the length of the coastline adjacent to the installation site boundary. The coastline length was estimated using a geospatial analysis from installation satellite images. Table C.5 lists the estimated length of coastline along the installation boundary used in the cost model.

Wildfire

The model to estimate the costs of improving installation resilience to wildfires was based on a community-level approach. The community-level resilience strategy included in our cost estimate is the creation of a defensible space, which is a natural or man-made area surrounding a community or property that has been treated to slow or stop the spread of wildfire. Resilience options for wildfires can also be implemented at a building level. Examples of building-level resilience strategies for wildfire include installing a Class A rated roof or upgrading exterior walls with ignition-resistant construction methods.[8]

---

[5] FEMA, *Requirements for the Design and Certification of Dry Floodproofed Non-Residential and Mixed-Use Buildings*, National Flood Insurance Program, Technical Bulletin 3, January 2021a.

[6] FEMA, 2007, Table 7-2; Jeroen Aerts, "A Review of Cost Estimates for Flood Adaptation," *Water*, November 13, 2018.

[7] We estimated unit costs for a 9-foot floodwall using a regression model because of the limitations of floodwall protection level options available in Hazus. The linear regression model using the five data points had an R2 value of 0.983.

[8] FEMA, *Wildfire Hazard Mitigation Handbook for Public Facilities*, FEMA P-754, October 2008.

TABLE C.4
Floodwall Unit Cost

Floodwall Protection Height	Unit Cost ($ millions)	Data Source
Floodwall (2-foot protection)	0.45/km	FEMA, 2007
Floodwall (4-foot protection)	0.69/km	FEMA, 2007
Floodwall (6-foot protection)	0.96/km	FEMA, 2007
T-Wall floodwall (13-foot protection)	15.7/km	Aerts, 2018
T-Wall floodwall (24-foot protection)	33.3/km	Aerts, 2018
Floodwall (9-foot protection)	7.8/km	Output from regression model

TABLE C.5
Floodwall Estimates, by Installation

Installation	Length of Coastline
NAVSTA Norfolk	11.3 km
NWS Yorktown	5.8 km
NAVSUPPACT Norfolk NSY	2.5 km
NAVSUPPACT Hampton Roads	0 km
Langley AFB	10.5 km
JBLE	21.2 km

Vegetation Management

The defensible space we assumed in our cost model for wildfire community-level protection is vegetation management around the installation site. The data source used to estimate costs for vegetation treatments is based on a unit cost per acre, included in a report by the Nature Conservancy.[9] We estimated the average cost for wildfire resilience treatments to be $1,000 per acre treated, which includes removal of vegetation that can fuel wildfire and utilization of a controlled burn. We considered other data sources for vegetation management costs, including FEMA Hazard Mitigation Project data in the OpenFEMA database; however, the data were reported as a lump sum, which made it difficult to determine a unit cost per acre.

To estimate the number of acres requiring treatment, we used satellite images of the installation sites from Google Earth and conducted geospatial analysis to measure the number of acres to treat. Our estimation of the number of acres requiring treatment is based on calculating the area of vegetation 1,000 feet from the

---

[9] Cecilia Clavet, Meryl Harrell, Rick Healy, Patrick Holmes, Chris Topik, David Wear, *Wildfire Resilience Funding: Building Blocks for a Paradigm Shift*, Nature Conservancy, June 2021.

FIGURE C.1

Cheyenne Mountain Space Force Station Defensible Space Area Estimation

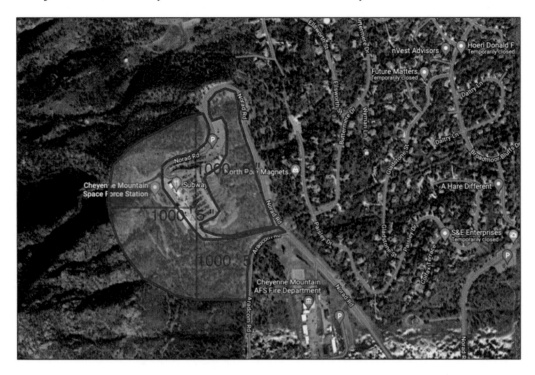

FIGURE C.2

Fort Carson Defensible Space Area Estimation

FIGURE C.3

Peterson Space Force Base Defensible Space Area Estimation

FIGURE C.4

Schriever Space Force Base Defensible Space Area Estimation

FIGURE C.5.

U.S. Air Force Academy Defensible Space Area Estimation

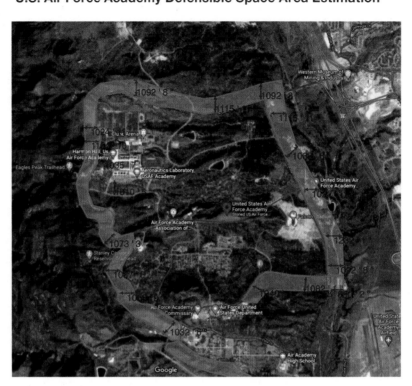

installation boundary, excluding any urban areas. Figures C.1–Figure C.5 illustrate the areas assumed to be part of the defensible space in our cost estimation model.

Cost Factors and Annualized Costs

After calculating the extended costs for each resilience strategy, we added cost factors to account for items, including general contractor fees, design fees, and contingencies (Table C.6). The method of adding cost factors and the percentage values for the cost factors are based on FEMA CEF guidance, which is used for post-disaster reconstruction or repair projects.[10]

Total costs with the cost factors were annualized for a given resilience option using a discount rate of 2.25 percent, which is based on discount rates used for federal water resource planning projects.[11] The useful life of each resilience strategy is based on FEMA literature and literature from other sources, as listed in Table C.7. We were unable to identify a useful life to assign for vegetation management because this depends on the frequency and likelihood of wildfire risk. As a conservative approach, we assigned a useful life of one year for vegetation management.

---

[10] FEMA, *CEF for Large Projects Instructional Guide*, ver. 2.1, September 2009.

[11] Bureau of Reclamation, Department of the Interior, "Change in Discount Rate for Water Resources Planning," *Federal Register*, Vol. 85, No. 239, December 11, 2020.

TABLE C.6
Factors Added to Cost Estimate

Cost Factor	Description	Added Percentage
General requirements	Includes additional costs for factors, such as safety and security, temporary services and utilities, quality control, and construction submittals.	8
General conditions	Represents a contractor's project management costs, including field supervision.	4
Design fees and contingencies	Accounts for design costs and unforeseen design conditions or circumstances.	20
General contractor markup	Includes general contractor overhead and profit (e.g., contractor office expenses, labor, and salary costs for personnel).	11
Other contingencies	Accounts for unexpected costs during the construction process.	8

TABLE C.7
Useful Life of Resilience Strategy

Resilience Strategy	Useful Life	Source
Roofing replacement	30 years	Roofing contractor guide[a]
Hurricane shutter installation	30 years	Unable to find literature on storm shutter useful life so we assume that useful life is same as roof replacement.
Dry floodproofing	22.5 years	Table 1 of FEMA P-1037[b]
Floodwall	50 years	Table 1 of FEMA P-1037[c]
Vegetation management	1 year	N/A[d]

[a] RGB Construction, "What Is the Average Lifespan of a Commercial Roof," webpage, undated.

[b] FEMA, *Reducing Flood Risk to Residential Buildings That Cannot Be Elevated*, FEMA P-1037, September 2015.

[c] FEMA, 2015, Table 1.

[d] Authors' estimate.

Abbreviations

AFB	Air Force Base
AGL	above ground level
CAPE	Office of Cost Assessment and Program Evaluation
CCI	city cost index
CECBH	concrete, engineered commercial building, high-rise
CECBL	concrete, engineered commercial building, low-rise
CEF	cost estimating format
DEM	digital elevation map
DoD	U.S. Department of Defense
DoDI	Department of Defense Instruction
EAP	expected annual probability
ESTCP	Environmental Security Technology Certification Program
FEMA	Federal Emergency Management Agency
GIS	geographic information system
JBLE	Joint Base Langley-Eustis
LEED	Leadership in Energy and Environmental Design
MOM	Maximum of Maximum Envelope of Open Water
MTBS	Monitoring Trends in Burn Severity
NAVSTA	Naval Station
NAVSUPPACT	naval support activity
NOAA	National Oceanic and Atmospheric Administration
NSY	naval shipyard
NWS	Naval Weapons Station
OSD	Office of the Secretary of Defense
PRV	plant replacement value
RPAD	Real Property Assets Database
RPUID	Real Property Unique Identifier
SERDP	Strategic Environmental Research and Development Program
SFE	square foot estimator
SLOSH	Sea, Lake, and Overland Surges from Hurricanes
USFS	U.S. Forest Service

References

Aerts, Jeroen, "A Review of Cost Estimates for Flood Adaptation," *Water*, November 13, 2018.

Air Force Civil Engineering Center, "Environmental Management," webpage, undated. As of October 27, 2022:
https://www.afcec.af.mil/Home/Environment/

Army Regulation 200-1, *Environmental Quality: Environmental Protection and Enhancement*, Department of the Army, December 13, 2007. As of October 27, 2022:
https://armypubs.army.mil/epubs/DR_pubs/DR_a/pdf/web/r200_1.pdf

Baldwin, Alexander J., *Developing Infrastructure Adaptation Pathways to Combat Hurricane Intensification: A Coupled Storm Simulation and Economic Modeling Framework for Coastal Installations*, thesis, Air Force Institute of Technology, March 2021.

Bureau of Reclamation, Department of the Interior, "Change in Discount Rate for Water Resources Planning," *Federal Register*, Vol. 85, No. 239, December 11, 2020.

Childs, Jan Wesner, "Rebuilding of Tyndall, Offutt Air Force Bases After Storms Stalled Due to Lack of Funds," Weather Channel, May 1, 2019. As of July 27, 2022:
https://weather.com/news/news/2019-05-01-tyndall-offutt-hurricane-flood-rebuilding-stalled

Clavet, Cecilia, Meryl Harrell, Rick Healy, Patrick Holmes, Chris Topik, and David Wear, *Wildfire Resilience Funding: Building Blocks for a Paradigm Shift*, Nature Conservancy, June 2021.

Colorado State Forest Service, "Protect Your Home & Property from Wildfire," webpage, undated. As of July 17, 2022:
https://csfs.colostate.edu/wildfire-mitigation/protect-your-home-property-from-wildfire/

Commander Navy Region Northwest, "Environmental Stewardship and Compliance," webpage, undated. As of October 27, 2022:
https://cnrnw.cnic.navy.mil/Operations-and-Management/Environmental-Stewardship-and-Compliance/

Copp, Tara, "71 of the Navy's T-45 Trainer Jets Could Not Evacuate Storm," *Navy Times*, August 25, 2017.

Dennis, F. C., "Fire-Resistant Landscaping," Colorado State University Extension, Fact Sheet No. 6.303, undated. As of July 17, 2022:
https://extension.colostate.edu/topic-areas/natural-resources/fire-resistant-landscaping-6-303/

Department of the Air Force, *Department of the Air Force Climate Action Plan*, October 2022.

Department of Defense Instruction No. 4165.14, *Real Property Inventory (RPI) and Forecasting*, Department of Defense, January 17, 2014, incorporating change 2, August 31, 2018. As of October 17:
https://www.esd.whs.mil/Portals/54/Documents/DD/issuances/dodi/416514p.pdf?ver=2018-12-18-095339-407

Department of Defense Strategic Environmental Research and Development Program and Environmental Security Technology Certification Program, "Regionalized Sea Level Change Scenarios," database, undated. As of July 27:
https://drsl.serdp-estcp.org/sealevelrise/562/feet

Dillon, Gregory K., and Julie W. Gilbertson-Day, "Wildfire Hazard Potential for the United States (270-m)," 3rd ed., Forest Service Research Data Archive, 2020. As of July 17, 2022:
https://www.fs.usda.gov/rds/archive/Catalog/RDS-2015-0047-3

DoD—*See* U.S. Department of Defense.

DoDI—*See* U.S. Department of Defense Instruction.

DoD Environment, Safety & Occupational Health Network and Information Exchange, "2021 Secretary of Defense Environmental Awards," webpage, undated. As of October 27, 2022:
found at https://www.denix.osd.mil/awards/2021secdef/

ESTCP—*See* Department of Defense Strategic Environmental Research and Development Program and Environmental Security Technology Certification Program.

Federal Emergency Management Agency, "Hazus," webpage, undated-a. As of July 10, 2022:
https://www.fema.gov/flood-maps/products-tools/hazus

Federal Emergency Management Agency, "Hazus User & Technical Manuals," webpage, undated-b. As of July 10, 2022:
https://www.fema.gov/flood-maps/tools-resources/flood-map-products/hazus/user-technical-manuals

Federal Emergency Management Agency, *Selecting Appropriate Mitigation Measures for Floodprone Structures*, FEMA 551, March 2007.

Federal Emergency Management Agency, *Wildfire Hazard Mitigation Handbook for Public Facilities*, FEMA P-754, October 2008.

Federal Emergency Management Agency, *CEF for Large Projects Instructional Guide*, ver. 2.1, September 2009.

Federal Emergency Management Agency, *Reducing Flood Risk to Residential Buildings That Cannot Be Elevated*, FEMA P-1037, September 2015.

Federal Emergency Management Agency, *Hazus Comprehensive Data Management System (CDMS) User Guidance*, April 2019.

Federal Emergency Management Agency, *Requirements for the Design and Certification of Dry Floodproofed Non-Residential and Mixed-Use Buildings*, National Flood Insurance Program Technical Bulletin 3, January 2021a.

Federal Emergency Management Agency, *Hazus Inventory Technical Manual: Hazus 4.2 Service Pack 3*, February 2021b.

Federal Emergency Management Agency, *Hazus 5.1 Flood Model User Guidance*, April 2022a.

Federal Emergency Management Agency, *Hazus 5.1 Hurricane Model User Guidance*, April 2022b.

FEMA—*See* Federal Emergency Management Agency.

Gordian, "RSMeans Data," webpage, undated. As of July 28, 2022:
https://www.rsmeans.com/

Hadley, Greg, "Tyndall and Offutt Rebuilds Need More Funds; Projected End Date in 2027–'28," *Air & Space Forces Magazine*, May 2, 2022.

Insurance Institute for Business & Home Safety, "Wildfire," webpage, undated. As of July 17, 2022:
https://ibhs.org/risk-research/wildfire/

Kenney, Caitlin M., "Air Force, Marine Corps Warn of Critical Readiness Concerns Without Supplemental Disaster Funding," *Stars and Stripes*, April 30, 2019.

Knutson, Tom, "Global Warming and Hurricanes: An Overview of Current Research Results," Geophysical Fluid Dynamics Laboratory, last updated February 9, 2023. As of July 27, 2022:
https://www.gfdl.noaa.gov/global-warming-and-hurricanes/

Kossin, James P., Timothy Hall, Thomas Knutson, Kenneth E. Kunkel, Robert J. Trapp, Duane E. Waliser, and Michael F. Wehner, "Extreme Storms," in Donald J. Wuebbles, David W. Fahey, and Kathy A. Hibbard, eds., *Climate Science Special Report: Fourth National Climate Assessment*, Volume I, U.S. Global Change Research Program, 2017.

Lindsey, Rebecca, "Climate Change: Global Sea Level," National Oceanic and Atmospheric Administration Climate.gov, April 19, 2022.

Monitoring Trends in Burn Severity, "Direct Download," datasets, undated. As of July 17, 2022:
https://www.mtbs.gov/direct-download

MTBS—*See* Monitoring Trends in Burn Severity.

Multi-Hazard Mitigation Council, *Natural Hazard Mitigation Saves: 2019 Report*, National Institute of Building Sciences, December 2019.

Narayanan, Anu, Debra Knopman, James D. Powers, Bryan Boling, Benjamin M. Miller, Patrick Mills, Kristin Van Abel, Katherine Anania, Blake Cignarella, and Connor P. Jackson, *Air Force Installation Energy Assurance: An Assessment Framework*, RAND Corporation, RR-2066-AF, 2017. As of January 31, 2023:
https://www.rand.org/pubs/research_reports/RR2066.html

Narayanan, Anu, Michael J. Lostumbo, Kristin Van Abel, Michael T. Wilson, Anna Jean Wirth, and Rahim Ali, *Grounded: An Enterprise-Wide Look at Department of the Air Force Installation Exposure to Natural Hazards: Implications for Infrastructure Investment Decisionmaking and Continuity of Operations Planning*, RAND Corporation, RR-A523-1, 2021. As of January 19, 2023:
https://www.rand.org/pubs/research_reports/RRA523-1.html

National Oceanic and Atmospheric Administration, "SLOSH Display Package," webpage, undated. As of May 4, 2022:
https://slosh.nws.noaa.gov/sdp/

National Hurricane Center and Central Pacific Hurricane Center, "Saffir-Simpson Hurricane Wind Scale," webpage, undated-a. As of July 16, 2022:
https://www.nhc.noaa.gov/aboutsshws.php

National Hurricane Center and Central Pacific Hurricane Center, "Storm Surge Maximum of the Maximum (MOM)," webpage, undated-b. As of January 24, 2023:
https://www.nhc.noaa.gov/surge/momOverview.php

Naval Facilities Engineering Systems Command, "NAVFAC Southeast Stands Up Resident Officer in Charge of Construction Sally for Hurricane Repair in Pensacola," September 16, 2022.

NOAA—*See* National Oceanic and Atmospheric Administration.

Office of the Assistant Secretary of the Army for Installations, Energy, and Environment, *United States Army Climate Strategy*, Department of the Army, February 2022.

Office of the Assistant Secretary of the Navy for Energy, Installations, and Environment, *Department of the Navy Climate Action 2030*, Department of the Navy, May 2022.

Office of Local Defense Community Cooperation, "Installation Resilience," webpage, undated. As of October 19, 2022:
https://oldcc.gov/our-programs/military-installation-sustainability

RGB Construction, "What Is the Average Lifespan of a Commercial Roof," webpage, undated. As of May 17, 2022:
https://rgbconstructionservices.com/average-lifespan-commercial-roof/

Rogoway, Tyler, "Setting the Record Straight on Why Fighter Jets Can't All Simply Fly Away to Escape Storms," *War Zone*, December 1, 2019.

Schriever Space Force Base, "3rd SOPS Backup Center Ready to Take On All Ops," undated.

Scott, Shirley V., and Shahedul Khan, "The Implications of Climate Change for the Military and for Conflict Prevention, Including Through Peace Missions," *Air and Space Power Journal–Africa and Francophonie*, 3rd Quarter, 2016.

SERDP—*See* Department of Defense Strategic Environmental Research and Development Program.

Short, Karen C., Mark A. Finney, Joe H. Scott, Julie W. Gilbertson-Day, and Isaac C. Grenfell, "Spatial Dataset of Probabilistic Wildfire Risk Components for the Conterminous United States," 1st ed., Forest Service Research Data Archive, 2016. As of July 17, 2022:
https://www.fs.usda.gov/rds/archive/Catalog/RDS-2016-0034

Stavros, E. Natasha, John T. Abatzoglou, Donald McKenzie, and Narasimhan K. Larkin, "Regional Projections of the Likelihood of Very Large Wildland Fires Under a Changing Climate in the Contiguous Western United States," *Climatic Change*, Vol. 126, 2014.

Sullivan, Becky, Noah Caldwell, and Ari Shapiro, "Nearly 8 Months After Hurricane Michael, Florida Panhandle Feels Left Behind," National Public Radio, May 31, 2019.

Sweet, William V., Radley Horton, Robert E. Kopp, Allegra N. LeGrande, and Anastasia Romanou, "Sea Level Rise," in Donald J. Wuebbles, David W. Fahey, and Kathy A. Hibbard, eds., *Climate Science Special Report: Fourth National Climate Assessment*, Volume I, U.S. Global Change Research Program, 2017.

U.S. Air Force, "Air Force Distributed Common Ground System," October 2015. As of July 27, 2022:

U.S. Army, "Army Quality of Life," webpage, undated. As of October 27, 2022:
https://www.army.mil/qualityoflife/

U.S. Army Installation Management Command, "Storm Damage," dataset, 2011–2021.

U.S. Department of Defense, *Base Structure Report–Fiscal Year 2018 Baseline: A Summary of the Real Property Inventory Data*, 2018.

U.S. Department of Defense, *Department of Defense Climate Adaptation Plan*, September 1, 2021.

U.S. Department of Defense, *Unified Facilities Criteria (UFC): DoD Facilities Pricing Guide*, UFC 3-701-01, March 17, 2022, incorporating change 1, July 15, 2022a.

U.S. Department of Defense, "DoD Announces Immediate and Long-Term Actions to Help Strengthen the Economic Security and Stability of Service Members and Their Families," news release, September 22, 2022b.

Vance, Kiara L., *Natural Infrastructure Alternatives Mitigate Hurricane-Driven Flood Vulnerability: Application to Tyndall Air Force Base*, thesis, Air Force Institute of Technology, March 2022.

Wuebbles, Donald J., David W. Fahey, and Kathy A. Hibbard, eds., *Climate Science Special Report: Fourth National Climate Assessment*, Vol I, U.S. Global Change Research Program, 2017.

Ziezulewicz, Geoff, "Hurricane Ian Prompts Evacuation of Navy Ships and Aircraft," *Navy Times*, September 27, 2022.